Oil Booms and Business Busts

Oil Booms and Business Busts

Why Resource Wealth Hurts Entrepreneurs in the Developing World

Nimah Mazaheri

OXFORD
UNIVERSITY PRESS

OXFORD
UNIVERSITY PRESS

Oxford University Press is a department of the University of Oxford. It furthers the University's objective of excellence in research, scholarship, and education by publishing worldwide. Oxford is a registered trade mark of Oxford University Press in the UK and certain other countries.

Published in the United States of America by Oxford University Press
198 Madison Avenue, New York, NY 10016, United States of America.

Library of Congress Cataloging-in-Publication Data
Names: Mazaheri, Nimah, author.
Title: Oil booms and business busts: why resource wealth hurts entrepreneurs in the developing world / Nimah Mazaheri.
Description: New York, NY: Oxford University Press, [2016] |
Includes bibliographical references and index.
Identifiers: LCCN 2015044613 (print) | LCCN 2015049345 (ebook) |
ISBN 978-0-19-049021-8 (hardcover) | ISBN 978-0-19-049022-5 (Updf)
Subjects: LCSH: Natural resources—Political aspects—Developing countries. |
Petroleum industry and trade—Political aspects—Developing countries. |
Entrepreneurship—Political aspects—Developing countries. | Economic development—Political aspects—Developing countries. | Industrial policy—Developing countries.
Classification: LCC HC59.7.M3566 2016 (print) | LCC HC59.7 (ebook) |
DDC 338/.04091724—dc23
LC record available at http://lccn.loc.gov/2015044613

Oil Booms and Business Busts

Oil Booms and Business Busts

Why Resource Wealth Hurts Entrepreneurs in the Developing World

Nimah Mazaheri

OXFORD
UNIVERSITY PRESS

OXFORD
UNIVERSITY PRESS

Oxford University Press is a department of the University of Oxford. It furthers the University's objective of excellence in research, scholarship, and education by publishing worldwide. Oxford is a registered trade mark of Oxford University Press in the UK and certain other countries.

Published in the United States of America by Oxford University Press
198 Madison Avenue, New York, NY 10016, United States of America.

Library of Congress Cataloging-in-Publication Data
Names: Mazaheri, Nimah, author.
Title: Oil booms and business busts: why resource wealth hurts entrepreneurs in the developing world / Nimah Mazaheri.
Description: New York, NY: Oxford University Press, [2016] |
Includes bibliographical references and index.
Identifiers: LCCN 2015044613 (print) | LCCN 2015049345 (ebook) |
ISBN 978-0-19-049021-8 (hardcover) | ISBN 978-0-19-049022-5 (Updf)
Subjects: LCSH: Natural resources—Political aspects—Developing countries. |
Petroleum industry and trade—Political aspects—Developing countries. |
Entrepreneurship—Political aspects—Developing countries. | Economic development—
Political aspects—Developing countries. | Industrial policy—Developing countries.
Classification: LCC HC59.7.M3566 2016 (print) | LCC HC59.7 (ebook) |
DDC 338/.04091724—dc23
LC record available at http://lccn.loc.gov/2015044613

CONTENTS

Acknowledgments *vii*

1. Introduction *1*
2. The Importance of *Everything but Oil* *17*
3. A Theory about Oil, Policy Makers, and the Business Environment *44*
4. Oil and the Business Environment in Iran *58*
5. Doing Business in India's Mining Belt *100*
6. Reform in the Gulf Arab Region: Saudi Arabia since the 1990s *129*
7. Conclusion *145*

Appendix: List of Interviews Cited in Chapter 5 *155*
Notes *159*
Bibliography *181*
Index *203*

ACKNOWLEDGMENTS

I benefited from the help of many people while writing this book. At the University of Washington, I was lucky to have Margaret Levi as my adviser, as she was a constant source of encouragement and insight. She motivated me to acquire the intellectual and methodological skills needed to undertake this project, for which I am extremely grateful. I also considerably benefited from the advice of Erik Wibbels. Even after he moved to Duke University, Erik was always there for me and helped me work through every aspect of this project. Ellis Goldberg was an important source of knowledge on the Middle East, and his work continues to influence me in numerous ways. Victor Menaldo was exceptionally generous with his time and broadened my understanding of the political economy of oil.

So much of what appears here came out of discussions during a book manuscript workshop held in 2013 and generously attended by Michael Ross, Benjamin Smith, Erika Weinthal, and Chris Heurlin. My admiration for Michael's work is clear in the pages that follow. His work directly paved the way for many people like me to embark on our own studies about the effects of oil. Erika's book on resource ownership had a big impact on me when I was in graduate school, and her encouraging words during a visit to Seattle in 2007 motivated me to take up this project. Ben deserves special thanks. He read a revised version of the manuscript and gave me timely advice on a number of occasions. I admire Ben greatly: he asks novel questions, can navigate doing research in different parts of the world, and can utilize different methodologies. Chris is a good friend and someone who I have turned to many times over the last decade for critical feedback. I would also like to thank David Art for taking the time to read the manuscript and provide detailed comments on every chapter. Thanks also to Kelly Greenhill and Ioannis Evrigenis for their advice on the book publishing process.

Since 2011, I have been fortunate to call Tufts University my intellectual home. I work in a wonderful department and am surrounded by helpful colleagues. Princeton University's Niehaus Center for Globalization and Governance kindly gave me a fellowship for 2013–2014, during which I expanded my skill set as a social scientist and received invaluable feedback. In 2011, I received a fellowship from Harvard University's Belfer Center for Science and International Affairs, which gave me resources and an exciting place to do work. Rick van der Ploeg and Tony Venables allowed me to spend a few months at the Oxford Centre for the Analysis of Resource Rich Economies in 2014, which was a marvelous experience. Early drafts of the chapters were written while working at the World Bank in Washington, D.C. from 2010–2011. I thank my colleagues there for providing me with a stimulating work environment and offering a sounding board for this project. The Gulf Research Center in Dubai graciously hosted me in 2009. Thanks also to staff members and faculties at the Empirical Implications of Theoretical Models institutes at the University of Michigan and Washington University.

In addition to the people mentioned, I benefited from discussions with and/or feedback from Chris Adolph, Faisal Ahmed, John Ahlquist, Benjamin S. Barber, Drusilla Brown, John Buchanan, Consuelo Cruz, Rob Devigne, Steffen Hertog, Phil Howard, Yannis Ioannides, Stephen Kosack, Homa Katouzian, Beth Kier, Pauline Jones Luong, Sunila Kale, Azza Layton, Michael McIntyre, Natalie Masuoka, Joel Migdal, Malik Mufti, Amit Prakash, Aseem Prakash, Dennis Rasmussen, Elizabeth Remick, Pearl Robinson, Deborah Schildkraut, Oxana Shevel, Tony Smith, Ashutosh Varshney, Mike Ward, Meredith Weiss, Susan Whiting, Matt Winters, and Eckart Woertz. Alisha Sett provided excellent research assistance and advice on India. Thanks also to Paula Driscoll for all her work on my behalf. I am grateful to the many people in India, Iran, and the United Arab Emirates who were generous with their time and patiently answered all of my (often incredibly naïve) questions.

At Oxford University Press, I have had the privilege of working with a talented Editor, Angela Chnapko. From the start, she has been enthusiastic about the project and I feel lucky to have had the opportunity to work with her. Princess Ikatekit, Jeyashree Ramamoorthy, and Christine Dunn have provided first-rate editorial assistance.

I am fortunate to have a wonderful partner, caring parents, and a thoughtful brother who value the importance of asking big questions about the world in which we live. It would take another book to truly express how I feel about each of them. My relatives and family friends—scattered around the world from North America to the Middle East—have been an endless source of encouragement. This book is dedicated, with love and admiration, to my father.

Oil Booms and Business Busts

Oil Booms and Business Busts

CHAPTER 1

Introduction

Starting a business and running it smoothly is a major challenge for entrepreneurs in oil-producing countries. In 2009, the *Financial Times* wrote about Iran's "deteriorating" business environment, where a 40 percent decline occurred in the number of permits awarded to private firms.[1] A professor at a business school in Nigeria told *Forbes* that if an entrepreneur wanted to transition from small-scale trading to manufacturing it is a "very tough road." He continued, "When regulation comes, it comes hard, and we feel the pain a lot more."[2] In 2010, it was estimated that 25 percent of the nine hundred thousand people in Russia's jails are accountants, entrepreneurs, legal advisers, or mid-level firm managers. A firm owner said, "The current environment is like swimming with crocodiles in a pool of sulfuric acid. . . . There's a war on business people in Russia."[3]

While oil-producing countries are frequently cited for their authoritarian tendencies,[4] it seems odd to find them excessively regulating entrepreneurs given what we know about their economic experiences throughout the last half century.[5] Despite their tremendous financial advantages since the oil boom era of the 1970s, most oil-producing countries have had unimpressive economic performance.[6] Much of this is attributed to the stagnation of *nonoil* sectors, such as manufacturing, which some describe as an inevitable feature of an oil-based economy. So why would oil producers overregulate the business community and repress local entrepreneurs, making it more difficult for key sectors like manufacturing to take root and grow? Why make things harder than they already are?

The reality is that oil-producing countries are much more likely than oil-poor countries to impose a host of barriers to the activities of domestic firms and entrepreneurs. In this book, I introduce this important finding and analyze it using statistical methods and case studies. I show that oil producers adopt policies that make it more difficult for entrepreneurs to start a business, acquire permits, pay taxes, and resolve contracts. I show that credit lending is lower in oil producers, something that is surprising given that an oil boom generally leads to an expansion of the financial sector. Managers of existing firms who are located in oil-producing countries are more likely to report spending more time dealing with government regulation and resorting to bribery in order to get around regulation. These findings hold irrespective of whether the oil producer is rich or poor, a democracy or autocracy, politically stable or unstable. Although some Gulf Arab oil states[7] emerge as weak outliers to this trend, I discuss in detail how their reforms are recent and tentative. Overall, oil wealth is strongly associated with government policy making that fosters a restrictive business environment.

This is a serious problem for these countries. A large body of research links the nature of a country's business environment—that is, the policies that a government adopts toward the private sector that affect the activities of domestic firms and entrepreneurs[8]—to a wide range of economic outcomes we care about. To be sure, these are outcomes that *all* countries care about, but that oil producers should *particularly* care about. We know that a business environment that fosters competition and facilitates commercial transactions is positively associated with the rate of economic growth.[9] Many link "pro-entrepreneur" policy making with the development of high-technology sectors, faster rates of innovation, reduced unemployment, and more.[10] In European countries, higher regulatory barriers depress the entry rate of new firms and cause existing firms to grow more slowly.[11] Any country seeking to pursue development through market-based growth needs to focus on the policies that define its business environment for domestic firms and entrepreneurs.

This book centers on the experiences of oil producers in the developing world,[12] and specifically how the business environment affects nonelites in the private sector. I use the term *nonelites* to refer to entrepreneurs of modest capital and owners of small- and medium-sized enterprises (SMEs).[13] I examine domestically-owned, privately-owned, nonexporting firms that conduct nonagricultural activities, such as industry or services.[14] Because I center on the developing world, I am naturally also interested in entrepreneurs and firms in the "informal sector,"[15] and those who occupy the space between the informal and formal sectors.[16]

To date, the writings on oil producers have largely ignored SMEs and nonelite entrepreneurs to focus instead on large firms and business elites. But the reality is that SMEs and nonelites conduct the bulk of economic activities and absorb a huge share of the labor force in these countries—as they do in most countries around the world. Since the 1970s, the majority of economic activities and new job creation has moved away from large firms to smaller ones.[17] Even if we ignore the informal sector, SMEs account for as much as 45 percent of employment and 33 percent of gross domestic product (GDP) in developing countries.[18] In some developing countries, such as Indonesia, firms that are categorized as micro-, small-, or medium-sized account for more than 99 percent of all firms and 97 percent of employment.[19] The importance of SMEs extends beyond the developing world. In the United States, nearly 90 percent of all firms have fewer than twenty workers and 63 percent of new jobs created from 1993 to 2013 were at small firms.[20] Although researchers continue to debate whether or not SMEs can serve as an "engine of economic growth,"[21] there is little doubt as to the central role they currently play in many economies around the world. And as I discuss in the following text, the business environment that SMEs face is an important indicator of a country's prospects for economic diversification and innovation.

An analysis of the business environment in oil producers can tell us a great deal about the state of entrepreneurship in these countries, which is often argued to be *the* critical factor in capitalist development.[22] Most of the writings that adopt this view take inspiration from Joseph Schumpeter, who believed that an evolutionary process takes place in capitalist economies.[23] Schumpeter saw technological progress at the heart of this process, which emerges endogenously through acts of innovation from entrepreneurial individuals. The process of innovation and technological progress is thought to drive economic growth over the long run. In countries both rich and poor, SMEs play an important role by fostering entrepreneurship and serving as a site of innovative activities.[24] The story of California's Silicon Valley perhaps best illustrates this phenomenon, as many of the now-leading firms in high-tech sectors have humble origins as entrepreneurial start-ups. But even in poor countries, research has highlighted the value and importance of entrepreneurship in shaping a range of social and developmental outcomes.[25]

Some research looks at the role of SMEs in developing countries through an analysis of the "missing middle" phenomenon, which occurs when countries are dominated by small- and large-sized firms but have comparatively few medium-sized firms.[26] A "missing middle" is one reason why many developing countries are saddled with high unemployment rates and struggle to grow. An analysis of the business environment can help us better understand the "missing middle" phenomenon. Indeed, in every country, governments establish the regulations and procedural requirements needed to legally start a business. They can make this process easy and straightforward or they can make it difficult, time consuming, and expensive. Governments play an important role in credit lending through state-owned banks and by establishing the institutions that govern lending practices, which impacts the volume of lending.[27] Consequently, the role of government policy making may be one key explanation for why a "missing middle" phenomenon occurs and is sustained over time.

An analysis of the business environment in oil producers can tell us about the challenges that informal entrepreneurs and firms face in these countries, which tend to represent a large share of domestic economic activities across the developing world. The International Labour Organization estimates that 50 to 75 percent of all nonagricultural employment in the developing world is located in the informal sector.[28] In sub-Saharan African countries, nearly 40 percent of GDP comes from informal sources and even in high-income Organisation for Economic Co-operation and Development (OECD) countries this figure is still about 14 percent.[29] Some scholars, notably Hernando de Soto, view the informal sector as a potential wellspring of entrepreneurship and innovation. He argues that stiff government regulation and unequal access to credit strangles informal entrepreneurs and wastes their potential contributions to economic development.[30] Although critics argue that the informal sector is unlikely to make a substantial contribution to development,[31] much of the skepticism is in regard to micro-sized firms and subsistence-oriented enterprises run by individuals living in extreme poverty.[32] Yet, in every developing country, there are medium- and even large-sized informal firms that can offer a bigger potential contribution to development should they formalize and gain better access to credit. However, many of them choose not to because the government makes formalization difficult and does not provide any clear benefits to becoming a formal firm.

Beyond the developmental implications, an analysis of the business environment can tell us about politics and a country's prospects for democracy—an issue that has long concerned researchers who study oil producers.[33] The repression of nonelite entrepreneurs and SME owners can have overarching political repercussions for governmental stability and the likelihood of democratization. A large body of research ties the growth of a "middle class"—of which nonelite entrepreneurs and SME owners are an important part—to the potential for democratization.[34] In Europe, alliances between capitalists and workers were crucial to unseating the exclusive political interests of elites and paving the way for democracy.[35] Research on the developing world is more limited, although a study by Eva Bellin found that industrialists (who generally have more capital than the firms or entrepreneurs under investigation here) are "contingent democrats."[36] Other studies suggest that small-scale entrepreneurs are accommodating, if not supportive, of authoritarianism.[37] However, we have reason to suspect that SME owners and nonelites will be less resistant to the redistributive consequences of democratization than business elites, and we would think their weak social position will give them more incentives to push for democracy. To be sure, a number of unanswered questions remain. Nevertheless, achieving an understanding of the business environment for nonelites in oil producers—countries long associated with authoritarian rule—should reveal insights about how government policy making can shape a country's possible pathways to democracy.

An analysis of the business environment also tells us something about "modernization," a highly debated concept but essentially defined as a passive and gradual process by which nonelites play a role in democratization over time. While the empirical evidence of a "modernization effect" in oil producers is inconsistent,[38] there is good reason to believe that fostering local entrepreneurship and a vibrant SME sector will contribute to the social process that leads to political reform over time. Critical components of the modernization process—such as occupational specialization, female entrepreneurship, and rising income for lower classes—benefit from government policies that create a supportive business environment for entrepreneurs and small firms. The fact that oil producers have disproportionately more restrictive business environments may provide us with a clue into why the modernization process has not taken hold in many of these countries.

Despite all that we can learn by studying how oil producers approach their business environments, we cannot make sense of their policy-making decisions in this area using the theories we currently have. There are two main theories about oil producers that we can apply toward an understanding of the private sector: the "rentier state" and the "Dutch Disease" arguments. Both are used, in various contexts, to explain the so-called resource curse findings whereby oil wealth leads to authoritarianism and slow economic growth.[39] The rentier state argument holds that oil wealth leads governments to avoid taxation, rely on patronage to buy off citizens, and repress the political rights of social groups—thereby fostering authoritarianism and political stability.[40] From an economic and developmental perspective, it is argued that rentier states are inefficient with a weak administrative capacity, a corrupt civil service, and an absence of rule of law.[41] Although the rentier state argument is sometimes stretched to explain any and all negative outcomes in oil producers, at its core is a theory about *fiscal power*. Oil producers do not need to collect tax revenues because they receive a constant and massive flow of petrodollars. Consequently, they can focus on stabilizing their rule and ignore the institution-building activities that would be required to grow an economy that was not blessed with such "manna from heaven."

While the rentier state argument can carry considerable predictive power, it is difficult to apply a purely fiscal-based theory toward an explanation of why oil producers have restrictive business environments. First, consider the argument's assumption that the onset of oil wealth leads governments to largely forego attempts at generating revenue from other means, such as by taxing firms and individuals. For one, the observation that oil producers do not tax is likely to reveal something about the legacy of institution building that *preceded* the discovery of oil, a factor highlighted in work by Benjamin Smith.[42] For that matter, such an assumption is inaccurate. Oil producers may tax less than oil-poor states, but their tax rates are not zero and, as I show in statistical analysis and case studies in the following text, they tend to impose higher tax-related burdens on SMEs. Second, the rentier state argument's assumption that oil wealth gives governments the financial means to exercise complete control over all decision making contrasts with the more complex reality discussed in work by Kiren Aziz Chaudhry.[43] The ample evidence of "rent seeking" from elites in oil-producing countries (addressed in detail in the following text) reveals that an oil windfall does not mean that the government can simply ignore the interests of powerful segments of society, whether now or in the future. And some may wonder: Why would such a government even *want* to ignore the interests of powerful segments of society?

Two of the cornerstone implications of the rentier state argument are that oil wealth will lead to authoritarianism and greater political stability. Neither of these outcomes is sufficient in explaining why oil producers would correspondingly foster a restrictive business environment. As I show in the following text, the link between oil wealth and a restrictive business environment holds irrespective of whether the government is a democracy or an autocracy. My detailed case study of a robust democratic state, India, helps to demonstrate that regime type does not play a definitive role in shaping the business environment in the context of abundant resource wealth. Finally, the oil-fueled-autocracy argument fails to account for the important exceptions discussed in this book. A small group of *highly* authoritarian oil states in the Gulf region have initiated business environment reforms while most oil producers have not.

The rentier state argument also predicts that oil wealth will lead to greater political stability, which may lead us to expect a *less* restrictive business environment.[44] Greater political stability means that it should

be far easier for oil producers to adopt these types of reforms. Given that there are clear "losers" from business environment reforms—namely, the business elites who would favor a more restrictive business environment with less market competition—oil producers are well poised to compensate them in exchange for reforms. Alternatively, the rentier state logic would suggest that business environment reforms are just another way that oil producers buy off a sizeable portion of the citizenry: nonelites in the private sector. Based on what the rentier state literature suggests about political stability, we would expect oil producers to have a less restrictive business environment—which they do not. It is furthermore important to note that political stability's effect on the prospects of reforms can cut either way. Political stability may offer the time and space needed to build a coalition to support reforms, but it can also demotivate leaders and render them complacent that reforms are disruptive and ultimately unnecessary.[45]

Despite the limitations of the rentier state argument when applied to understanding the business environment in oil producers, its spotlight on rent seeking is a central part of the theory that I present. The concept of rent seeking, as applied in the work on natural resources, is the practice of pursuing special benefits and privileges from the government. The actual benefits are called "rents." Rent seeking occurs with great frequency in places where the government generates a significant amount of revenues from valuable natural resources, such as oil.[46] As unsavory as rent seeking sounds, there is often a shrewd logic at hand. Leaders trade rents for political support or quiescence from key elites, such as those with business interests. Correspondingly, elites back leaders while using rents to increase their personal wealth and conquer portions of the private sector.

The second main theory that applies to oil and the private sector is the "Dutch Disease" argument. The term *Dutch Disease* has its origins in the 1970s, when the Netherlands' manufacturing sector declined after the discovery of abundant natural gas fields. The term is broadly applied to countries that experience various economic ills as the result of oil dependence. The Dutch Disease argues that significant oil exports lead to overvaluation of a country's currency, whereby a country's exports become more expensive and imports become cheaper. Additionally, a factor movement effect takes place after an oil boom, whereby capital and labor shifts to the oil sector and nontradeable sectors. The upshot of the Dutch Disease is that nonoil exporters as well as producers for the domestic market become less competitive.[47] Research is often supportive of the Dutch Disease effect, but there are some inconsistent findings as well.[48] Public intellectuals and the media in oil-producing countries sometimes decry the Dutch Disease as the source of all their economic and social ills.[49]

Although the Dutch Disease argument has served as the traditional way to explain why oil producers end up with uncompetitive manufacturing sectors, it needs to be recognized that the causal story it offers is one that exclusively centers on macroeconomic factors. But we know that currency appreciation and shifts in capital and labor are not the only mechanisms that cause domestic firms to become uncompetitive over time—the government and its policies certainly have a role to play. Many agree with this statement. Jeffrey Sachs argues, "this fear [of the Dutch Disease] is vastly overblown if the oil proceeds are being properly invested as part of a national development strategy."[50] Terry Lynn Karl holds, "The Dutch Disease is not automatic. The extent to which it takes effect is the result largely of decision-making in the public realm."[51] Miriam Lowi argues, "Indeed, Dutch Disease is not inevitable. As long as national governments implement policies to rein in private consumption and investment expenditures, and adopt productive investment programs in the non-oil tradable sectors, its effects can be minimized, if not averted."[52]

And yet, when the policy-making component is discussed in writings about the Dutch Disease (which is rare), the focus is usually on government policies that mitigate the appreciation of the real exchange rate. Other government policies that directly affect the competitiveness of firms—such as the regulatory barriers that compose a country's business environment—are ignored. But even if the macroeconomic effects of Dutch Disease render domestic firms less competitive over time,[53] the government is not simply a passive observer in this process. The research on business and regulation demonstrates that policy making and the business environment play a key role in the expansion (or contraction) of economic sectors. To put it plainly, entrepreneurs and investors in manufacturing (or any sector for that matter) are unlikely to remain competitive over time if there are onerous and arbitrary regulatory barriers, no opportunities for credit, and predatory officials that rely on bribes from firms to subsidize their personal income. Neglecting the role of government policy making and the nature of a country's business environment may leave out a crucial component of why many domestic firms in oil-producing countries struggle to survive.

Related to the Dutch Disease argument are studies that employ formal economic models that explore how oil, or natural resources more generally, affects entrepreneurship and the preferences of entrepreneurs.[54] Most begin with the assumption that oil wealth should crowd-out entrepreneurial activities given a wage premium in resource-related and nontradeable sectors. Although data limitations restrict proper testing of these theories (especially in the developing world), a larger problem is that the vast majority of these studies do not address the role of the government or specific policies that support or inhibit entrepreneurship. Entrepreneurs are also frequently assumed to be a homogenous and static group of economic actors. In reality, there is significant variation across time in the wealth, political power, and class designation of entrepreneurs in any country. Those defined as nonelites one day may become elites the next. Recent work in political economy has underscored the differences between elites and nonelites during the process of economic development,[55] an important nuance that I highlight in my theory in the following text.

Because this book focuses on government policy making, it has a practical application in that it can offer policy recommendations for oil producers. Academic researchers and development practitioners are increasingly interested in uncovering which policies are important for shaping social, political, and economic outcomes in oil producers, and what factors determine whether or not these policies are pursued.[56] And yet, throughout the large body of work on the resource curse and what policy makers can do to mitigate its effects, the role of the business environment is either not addressed or it is only briefly mentioned.[57] This book provides the first comprehensive inquiry into the subject.

1.1 THE ARGUMENT

I explain how oil producers approach their business environments by focusing on how an oil windfall[58] shapes the *interests* of policy makers and business elites about the supply of reforms and the *capacity* of nonelites to demand reforms. I utilize the concepts of "rent seeking" and "rent seizing," and therefore build on the knowledge we already have about resource windfalls and the unique relationship that unfolds between government and business in the context of resource wealth.[59] In my theory, a conceptual distinction is made between *elites* and *nonelites* in the private sector. Nonelites are entrepreneurs of predominantly modest capital and owners of firms that generally correspond to an SME in a country. These individuals and units are different from business *elites*, who have more capital, tend to own larger firms, and are more likely to command political influence on an individual level.

The practice of rent seeking after an oil windfall conjures up an image that is familiar to us all. The stories told about "petrostates" depict this very phenomenon: the discovery of oil sparks the growth of a class of Ferrari-driving elites, who forge channels of privilege by courting policy makers and receiving special benefits from the government. While a select group of elites manage to directly siphon off cash from the oil sector, most receive indirect benefits such as lucrative government contracts, free lines of credit for their business ventures, and special advisory positions. To be sure, this story has been told in many contexts and time periods.

What becomes important for my argument is how business elites react to an oil windfall and correspondingly seek to protect their rents as a way of limiting competition with nonelites. All else equal, elites and nonelites should have opposing preferences about whether or not the government should pursue business environment reforms that lower the barriers to market entry, ease restrictions on firms' activities, improve access to credit, streamline the permit application process, and so forth. Nonelites should support business environment reforms, but elites should want to block them as a means of limiting competition and raising their profits. While this assumption is likely to hold irrespective of whether or not a country receives an oil windfall, I argue that an oil windfall *amplifies* the difference in preferences and, more importantly, *dampens* the interests of policy makers about adopting business environment reforms.

First, consider how an oil windfall affects the interests of elites about business environment reforms. Given that their primary motivation for seeking rents after an oil windfall is to receive preferential treatment and benefits from the government, it is rational for them to want to protect these rents and limit the extent to which nonelites compete with them. In effect, an oil windfall fosters its own windfall for business elites through the rent-seeking mechanism and protecting these rents should be their paramount concern. Uncertainty about the outcome of reforms should increase their efforts to block them, especially when nonelites pose a credible threat to their business activities and if the details of reforms are vague. Although we know that a highly restrictive business environment is an acute challenge to the business activities of nonelites, it has little effect on elites who receive rents. Unsurprisingly, "favored" firms use their political connections or bribery to get around these barriers.[60]

Next, consider how an oil windfall affects the interests of policy makers about business environment reforms. I argue that, all else equal, policy makers in countries that receive an oil windfall should be opposed to business environment reforms. Foregoing reforms will help to solidify a policy maker's relationship with business elites, and they should be eager to do this as elites can serve as an important source of political strength. The rent-seeking dynamic between policy makers and elites tightens their bond, and provides elites with an avenue to express their opposition to reforms. An oil windfall should reinforce a policy maker's motivation to strengthen their relationship with elites, as this bolsters their rule and allows them to continue their command over a lucrative oil sector and control the allocation of rents (termed *rent seizing*)—which carries its own personal benefits for policy makers.

Beyond the interests of policy makers and business elites, I argue that an oil windfall should blunt the capacity of nonelites in the private sector to demand reforms. A typical reaction to an oil windfall is that the government becomes completely focused on extracting the resource and boosting the growth of this industry alone. Albert Hirschman described this as the development of an "enclave economy,"[61] and, similar to the Dutch Disease, the outcome is that the nonoil portions of the economy are crowded out by a booming oil sector. I argue that this economic implication of an oil windfall has political implications for nonelites in nonoil sectors. The declining role of nonoil sectors on relative economic terms will weaken the capacity of nonelites in these sectors to influence policy makers. New entrepreneurs and nonelites are more likely to be underrepresented at the policy formation stage in countries with expanding oil sectors. Consequently, after an oil windfall, the possibility of a constructive dialogue between policy makers and nonelites should be low. The power of business associations who advocate on behalf of nonoil sectors or the specific interests of SMEs is likely to weaken over time.

Absent the capacity to formally demand reforms, nonelites in the private sector can exit the market, engage in collective action, or simply endure a restrictive business environment. For many domestic entrepreneurs and SME owners who are the focus of this book, exiting the market is infeasible as they have limited capital, they have few transferable skills, and migration is unrealistic. Collective action, although feasible, should have the opposite of its intended effect and alienate nonelites from policy makers. This will reduce the likelihood that nonelites are able to influence policy down the road. Enduring a restrictive business environment seems a more likely outcome. Over time, however, a restrictive business environment will discourage others to become entrepreneurs and will challenge the ability of existing firms to stay competitive in the market.

My theory focuses on the effects of oil windfalls, but there is little reason to assume that another type of resource windfall—such as coal, gold, or copper—will have a different effect. That being said, testing my theory in the context of nonoil resources is harder because few countries in the world receive nonoil windfalls that are anything close to what oil producers get. From 1970 to 2009, Australia was the world's top nonoil resource income earner (mostly coal, iron ore, gold, and other metals). Yet, over this same period, oil brought *nine times* as much income to Saudi Arabia and *three and a half times* as much income to Iran. If rent seeking increases with the magnitude of the windfall, as it theoretically should, we would expect oil producers to be particularly prone to this dynamic. In order to test my theory's applicability to nonoil resource windfalls, I take advantage of the unique history of state formation in India. I analyze the creation of Jharkhand, a coal-dependent state, as a way of testing my theory in a nonoil context, at the subnational level of policy making, and in a democratic state. All in all, I find little difference between an oil-producing country and a coal-producing state in terms of how the government approaches the business environment.

My theory provides a new way of thinking about government policy making after an oil windfall. By outlining a "logic of policy making" in oil producers, I contribute something that is rarely done with respect to these countries. Quite often, policy making in oil producers is described as "arbitrary, irrational, and volatile,"[62] or lacking coherency.[63] But we know that the stakes are incredibly high for oil producers to get policy making right. If we agree this is the case, there may be an underlying logic that helps explain why oil producers implement certain policies that seem from the outside to be irrational or illogical. Perhaps, from the standpoint of policy makers, they are not.

1.2 WHAT LIES AHEAD

The book consists of seven chapters. I utilize statistical and case-study methodologies to examine how oil wealth shapes government policy making toward the private sector over time, specifically among developing countries. In Chapter 2, I start by employing cross-national data in a descriptive manner to answer a fundamental question: How have oil producers fared in expanding their *nonoil incomes* throughout the last half century? This is my way of "setting the stage" because we first need to know the extent to which this is actually a problem for these countries. The data reveal that oil producers have encountered grave difficulties in growing their nonoil incomes over time. This observation is all the more remarkable given the advantage they once had over oil-poor countries, not to mention the huge amount of revenues they could have marshaled toward increasing their nonoil incomes.

Chapter 2 then employs statistical techniques for cross-national and panel data to inspect the link between oil and a country's business environment. I show that oil is associated with higher barriers and greater difficulties in starting a business, acquiring permits, paying taxes, and resolving contracts. These findings specifically speak to the business environment for domestic SMEs. I also show that oil is negatively associated with credit lending to the private sector. These relationships emerge after controlling for a range of variables and subjecting the analyses to robustness checks. Furthermore, they are not driven by the presence of Middle Eastern oil states in the sample of countries.

If oil producers regulate SMEs more, this should affect the behavior of firm owners and managers. Using statistical techniques, I reveal positive associations between oil and (1) the time that firm managers spend dealing with regulation and (2) the percentage of managers that report bribing public officials to get around regulation. Overregulation of SMEs in oil producers should also affect their rates of entrepreneurship and prospects for innovation. I provide results that address both of these areas, although it is important to clarify that I am not able to statistically test for a causal pathway that leads to these outcomes. Nevertheless, I am able to demonstrate that oil wealth is negatively associated with levels of entrepreneurship (new firm entry rates) and innovation (patent applications), after controlling for other plausible factors.

But not all oil producers have a restrictive business environment. A select few, mostly the Gulf Arab oil states, display tentative signs of adopting business environment reforms that most of their oil-producing colleagues have not adopted. While there is variation among this group and these countries are fundamentally *different* from most oil producers,[64] signs of reform render them weak outliers. Teasing out the forces that lead to reform is the goal of my analysis of Saudi Arabia in Chapter 6, given that it can offer clues into how other oil producers might pursue similar reforms down the road.

Chapter 3 outlines a probabilistic theory to explain the link between oil wealth and a restrictive business environment. In essence, I argue that an oil windfall dampens the interests of policy makers and business elites about the supply of business environment reforms. An oil windfall also blunts the capacity of nonelites to demand these reforms. The supply-side component is explained by how an oil windfall leads to an alignment of the political incentives of policy makers with business elites' goals of rent seeking in the context of uncertainty. I situate my theory within the existing literature and adopt a cost-benefit approach to generate supply-side predictions about how an oil windfall affects policy makers' interests. The demand-side component of my theory is explained as the political implication of the crowding-out of nonoil sectors after an oil windfall.

I then test my theory using a case-study approach. Although the statistical tests I employ are useful for isolating the relationship between oil and the business environment, they cannot assist with the task of causal inquiry. A case-study approach is needed to reveal causal processes, and I rely on the techniques of process tracing throughout each case-study chapter.[65] To be sure, case studies offer far richer evidence of government policy making than what is discernible from the cross-national quantitative measures analyzed in Chapter 2. They are critical for properly understanding the interplay among policy makers, business elites, and nonelites that is at the heart of my theory. Additionally, a case-study approach facilitates a test of my theory under different conditions in order to assess the scope of its applicability.

Chapter 4 opens the case-study section. My first case study is of Iran, one of the world's foremost and long-term oil producers. A study of Iran is justified for a number of reasons, one of which is that it lies "on the regression line" (discussed more in the following text) and is therefore an appropriate case for testing the theory. Equally important is the fact that Iran is different from other long-term oil producers in that its oil windfall spans two political eras that are separated by the 1979 revolution that deposed a monarchy. Thus, Iran is different from the Gulf Arab oil states because it has experienced a significant political transition since becoming an oil producer. Confirming an effect of oil wealth on Iran's business environment over two different political systems helps to build confidence that the theory holds on a broader level. First, I provide an overview of policy making in Iran prior to the oil windfall that resulted from the 1953 nationalization crisis. I then examine Iran's business environment between 1954 and the 1979 revolution, when the country was under the rule of Mohammad Reza Shah. Next, I center on the period following the 1979 revolution. I rely on primary and secondary source materials ranging from Iranian newspapers to time-series data sets on manufacturing licenses.

One of the important lessons from Chapter 4 is that the effect of oil wealth on policy making remains constant beyond the threshold of a revolution. During the rule of Mohammad Reza Shah, business elites served as his main source of support and he structured policy and economic institutions in order to exclusively benefit this group. His government simultaneously squeezed nonelites in the private sector on economic, political, and even physical terms. But after the 1979 revolution, the business elites who prospered under Mohammad Reza Shah were liquidated. However, a new group of elites emerged—*bonyads* (quasipublic foundations) and similar units—to take up their role with a similar effect on the business environment for nonelites.

An analysis of Iran offers many insights about the specific contours of policy making in a long-term oil producer, but the theory needs to be tested under different conditions and on cases that lie beyond our sample of oil producers in the world. Chapter 5 offers such a test through a comparison of two Indian federal states, Bihar and Jharkhand. These two states share a unique history in that they were part of the same federal unit until November 2000. The bifurcation of (undivided) Bihar left a formerly resource-rich Bihar now a resource-poor state, and created a highly resource-rich state named Jharkhand. Comparing these two states allows me to better isolate the effect of a nonoil resource windfall (primarily coal) and test the theory at the subnational level of policy making and in a democracy. The material used to construct my case comparison comes from secondary sources, economic data, newspapers and media reports, and interviews conducted in both states as well as in New Delhi.

Chapter 5 begins with an overview of the political economies of Bihar and Jharkhand since India's independence in 1947. I then compare the two states' business environments after the bifurcation in November 2000 and explain the divergence in policy making that takes place. I examine the case of Jharkhand in depth by focusing on how the coal windfall shaped the interests of policy makers and business elites, while weakening entrepreneurs and firms in nonmining sectors. I conclude that, despite their numerous differences, Middle Eastern oil producers and Indian coal producers adopt similar policies toward firms and entrepreneurs. I also show how resource-poor Bihar has adopted policies after the bifurcation that structure a more supportive business environment for entrepreneurs and SMEs, which is what my theory would predict.

What about the weak outliers identified in Chapter 2? Chapter 6 teases out the causal mechanisms that can explain how a select group of oil producers, namely the Gulf Arab oil states, have adopted limited business environment reforms. I analyze the case of Saudi Arabia, perhaps the country most associated with the word *oil*. I reveal the conditions under which business environment reforms emerged in Saudi Arabia, relying on secondary sources, economic data, and media reports. I concentrate on the role of a monarchical system and the institution of economic familism, facets of Saudi Arabia's political economy that are visible in the other Gulf Arab oil producers. I show how policy makers' continual reinforcement of economic familism reduces the uncertainty associated with business environment reforms, and the propensity for business elites to view new entrepreneurs as a threat to their interests.

Chapter 7 returns to my central argument that an analysis of policy making that defines the business environment gives us important insights into the challenges that oil producers have faced and tells us a great deal about where they are going. Since the start of the world's oil boom, oil producers have collectively failed to foster business environments that support the growth of domestic entrepreneurs and SMEs. Instead, they have centered on ferrying rents to business elites with whom policy makers have direct ties. Even when the group of elites was liquidated (as following the 1979 revolution in Iran), a new one was quickly fashioned. The effect this has on the growth of SMEs and the activities of nonelites in the private sector has been dramatic, and carries direct implications for our understanding of a range of social, political, and economic outcomes in these countries. The economist Paul Collier recently wrote that, "The failure to harness natural capital is the single-most important missed opportunity in economic development."[66] My book helps explain the roots of this missed opportunity, and provides ideas about how it can be avoided in the future.

CHAPTER 2
The Importance of *Everything but Oil*

There is no shortage of research that calls attention to the various problems that afflict natural resources producers, in particular oil producers. Many bookcases could be filled with writings that employ the term *resource curse* to refer to the negative political, economic, social, and conflict-related outcomes that are said to accompany abundant natural resources.[1] Philosophers such as Jean Bodin and Adam Smith posited arguments that invoke the concept of the resource curse.[2] Policy makers in oil-producing countries candidly speak of the ills that oil wealth brings their nation and people. One of the most famous quotes comes from Sheikh Ahmed Zaki Yamani, Minister of Petroleum and Mineral Resources for Saudi Arabia from 1962 to 1986, who reportedly said: "All in all, I wish we had discovered water instead."

My starting place is one of agnosticism regarding the consequences of oil wealth. I do not begin with an argument that oil *should* have a negative or positive effect—and then seek to confirm or establish the conditions under which this argument holds. Instead, I first seek to understand the extent to which oil producers have faced development-related challenges over time, specifically in how they have fared with respect to growing their *nonoil* sectors. I am principally interested in oil producers in the developing world, but I am also interested in the extent to which dynamics present in these countries might manifest in others who produce different types of natural resource wealth.

While agnostic about the consequences of oil wealth, I argue that an oil producer needs to consistently expand its nonoil sectors in order to prosper over the long term. Expanding the nonoil sectors is a way of decreasing vulnerability to economic shocks, increasing opportunities for local employment, and reducing the revenue constraints associated with oil dependence. Some of the existing research on oil overlooks the importance of the nonoil sector, while focusing more on what countries should do with oil revenues—how to manage and invest the revenues, how to save for the future, the issue of revenue transparency, and so forth.[3] To be completely clear, these issues are all very critical for oil producers to address. Yet, I argue in this book that oil producers need to be just as concerned about their *nonoil* income as they are about their oil income.

2.1 TRENDS IN NONOIL INCOME OVER TIME

I start with a purely descriptive account of trends in the nonoil incomes of all countries since 1960 as a way to understand what the world looks like, what has changed, and why this is a pressing problem for oil producers in particular. I construct a measure of a country's nonoil income per capita by subtracting the value of oil and natural gas production from its GDP in a given year (measured in constant 2000-year US dollars), and then divide this value by the total population.[4] Thus, if a country does not produce any oil or natural gas (what I refer to as "nonproducers"), the value is the exact same as their GDP per capita. If a country does generate income from oil production, the resulting value is *everything but oil*.[5] I calculate annual measures for 165 countries from 1960 to 2010.[6] I zero in on the experiences of the twenty-seven long-term oil producers in the developing world (refer to the list in Table 2.1). I use the definition established by Michael Ross, which categorizes a country as a long-term oil producer if it generates at least 100 US dollars per capita in oil and gas income for at least two-thirds of the time since 1960, or two-thirds of their sovereign years after independence.[7]

We know that long-term oil producers are critically reliant on their oil income to fund the government's activities and address the basic needs of the population. Particularly since the 1970s, the steady stream of oil revenues to these countries has led to an increase in national incomes, a boom in the development of the resource sector, and (in many cases) an expansion of the public provision of vital goods and services to the population. However, over this same time period, long-term oil producers have encountered serious difficulties in increasing their nonoil incomes. As a group, long-term oil producers in the developing world experienced a 23 percent decline in their average nonoil income throughout the last half century. And yet, throughout this same time period, nonproducers in the developing world made huge gains by increasing their average nonoil

Table 2.1 LIST OF DEVELOPING COUNTRIES THAT ARE OIL PRODUCERS IN 2010

Country	Oil income per capita in 2010	Country	Oil income per capita in 2010	Country	Oil income per capita in 2010
Algeria*	1680	Equatorial Guinea	11235	Russia*	2145
Angola*	2396	Gabon *	3917	Saudi Arabia*	7877
Argentina*	505	Indonesia	142	Sudan/South Sudan	349
Azerbaijan*	2985	Iran*	1640	Suriname	699
Bahrain*	2199	Iraq*	1783	Syria *	474
Belize	283	Kazakhstan*	2429	Thailand	151
Bolivia	269	Kuwait*	19814	Trinidad and Tobago*	5881
Brazil	256	Libya*	6551	Tunisia	211
Brunei*	11765	Malaysia*	789	Turkmenistan*	1945
Chad	258	Nigeria*	409	United Arab Emirates*	8708
Colombia	476	Oman*	8633	Uzbekistan*	314
Croatia	130	Papua New Guinea	104	Venezuela*	2020
Cuba	122	Peru	118	Yemen	417
East Timor	1755	Qatar*	22598		
Ecuador*	782	Republic of Congo*	1992		
Egypt	267	Romania*	157		

Per the methodology outlined in Ross, a country is defined as an *oil producer* in a given year if it has at least 100 US dollars per capita in income from oil and natural gas (measured in constant 2000-year US dollars). Countries identified as *long-term oil producers* are indicated with an asterisk (see Ross 2012, 16 and 20–22). Data are available from Ross 2013.

income by 169 percent (see Figures 2.1 and 2.2). Basically, those *without* oil did far better in expanding nonoil sectors than those *with* oil.

The world looked quite different before the first oil boom in October 1973. On the cusp of the boom, the average nonoil income of long-term oil producers was more than *six times higher* than all other developing countries. The Gulf Arab oil states skew this difference, but even when these countries are excluded the average nonoil income of long-term oil producers is still nearly three times higher than the rest of the developing world. However, the average nonoil income of long-term oil producers drops during the oil boom era that lasts from the fall of 1973 until

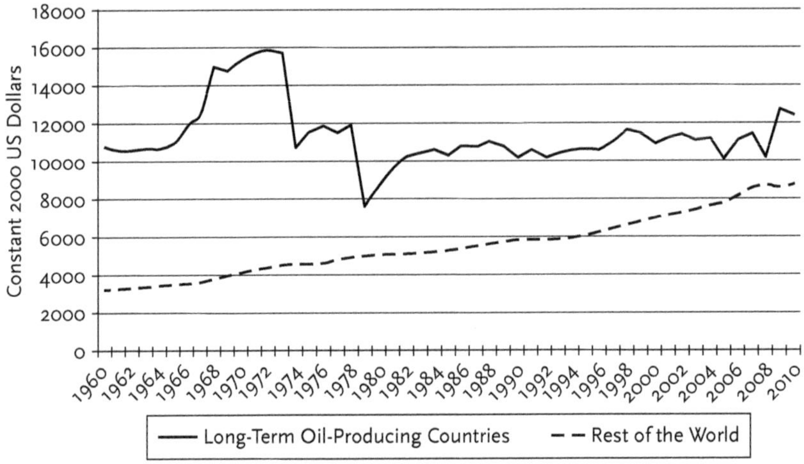

Figure 2.1 Average Income from Nonoil Activities, 1960–2010

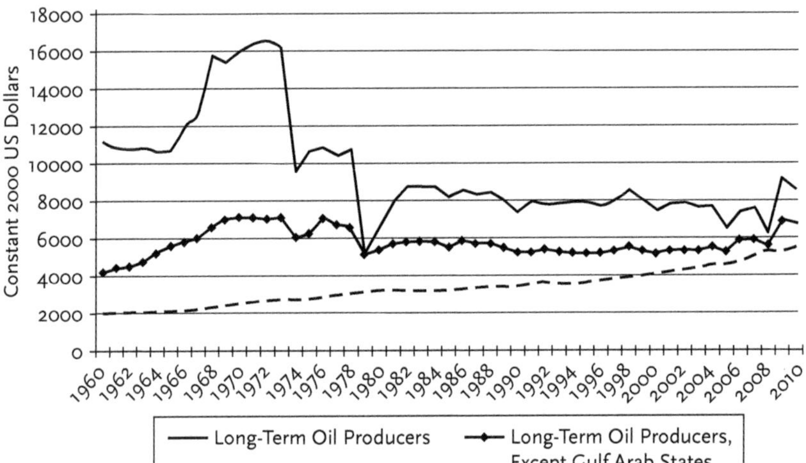

Figure 2.2 Average Income from Nonoil Activities, Developing World Only, 1960–2010

1985. Oil producers seem to never fully recover from the drop. During the oil bust era that then lasts from 1986 until 2003, we should have witnessed oil producers furiously pushing for the expansion of their nonoil sectors. It seems logical that they would now try to expand other parts of their economies because high oil prices appeared to be a thing of the past. Crude oil prices declined from an average of ninety-six dollars a barrel in 1980 to less than thirty dollars a barrel by 1986. Prices remained below forty dollars a barrel until 2004.[8] But as Figure 2.2 illustrates, the average nonoil income of long-term oil producers actually declined 10 percent during the oil bust.

Anyone looking at these trajectories would guess that the average nonoil income for nonproducers will surpass oil producers in the near future. Although some signs of a rebound from oil producers appear after 2008, it is unknown—and unlikely given the historical trend—if this will continue. The steady decline in the nonoil income of oil producers is all the more extraordinary given the substantial advantage they once had over the nonproducers, not to mention the huge amount of revenues they *could* have used toward boosting their nonoil sectors. Indeed, Figure 2.2 is a striking reminder of what Paul Collier called the "single-most important missed opportunity in economic development" (Collier 2010, 37).

The difficulties that long-term oil producers experience in expanding their nonoil incomes remain one of the most urgent issues for these countries to address. I argue that we can better understand the roots of this trend—as well as generate ideas about how to forestall it—through an investigation of the policies that oil producers adopt toward domestic firms, entrepreneurs, and nonoil sectors. Let me be clear: I am not arguing that Dutch Disease, inefficient state-owned enterprises, or the mismanagement of oil revenues plays no role in the stagnation of nonoil sectors in these countries throughout the last fifty years. All of these factors matter. But while we already have a good idea about the role these factors play, we have little knowledge about policy making toward the private sector and the business environment for entrepreneurs and SMEs. Next, I start by examining the relationship between oil wealth and the nature of regulatory barriers in the business environment.

2.2 OIL, REGULATION, AND THE BUSINESS ENVIRONMENT

Oil producers consistently impose higher regulatory barriers on the activities of domestic entrepreneurs and SMEs, specifically as they relate to firm start-up, obtaining permits, paying taxes, and resolving contracts. This is readily observed by examining data published by the World Bank's *Doing Business* project. The *Doing Business* project assembles the most comprehensive time-series cross-sectional data set on the business environment currently available (data for 140 countries are available starting in 2004; by 2010, data are available for 163 countries). It contains measures that specifically address policy making as it applies to the activities of entrepreneurs and domestically-owned, small-sized firms.[9] The types of firms analyzed in the data set closely align with the book's focus on SMEs and nonelite entrepreneurs. The firms possess the following characteristics:

> [the firm] performs general industrial or commercial activities; it does not participate in foreign trade; it is a domestically-owned limited liability company; it rents land and business premises; it has between five and 50 employees one month after the commencement of operations all of whom are nationals; it has turnover of up to 10 times its start-up capital; and it does not qualify for investment incentives.[10]

This allows for an examination of firms that are not likely to be elite owned but are still larger than microenterprises. They are seeking official entry and thus are either avoiding the informal sector or seeking to transition from the informal into the formal sector. Furthermore, the firms cannot be foreign owned, which allows me to get at *domestic* entrepreneurs. I utilize the measures in the data set that address entrepreneurship, the activities of firms defined in the preceding manner, and firms that are not export oriented, publicly traded, or have more than sixty employees.

We know that regulatory policy is a key component of a country's business environment, particularly in the way it affects entrepreneurship and the ease of opening a firm. Governments regulate who can or cannot legally open a firm; how easy or difficult is the process; and other considerations. Regulation of entrepreneurs entails a *procedural* component, whereby entrepreneurs must interact with government agencies, lawyers, auditors, or notaries to complete requirements. Common examples of regulatory procedures are to register for the income tax or present the name of the company for the government's approval. In 2012, two countries required just one procedure to open a firm: Canada and New Zealand. At the other end of the spectrum is Equatorial Guinea, where eighteen procedures are required. An initial analysis of the data set revealed that most regulatory procedures are not related to the government's concerns about taxation, labor, or the environmental effects of business activities—instead, they are a form of "screening."[11]

Regulation also entails a *time* component for entrepreneurs seeking start-up, whereby prospective owners wait as bureaucracies and relevant government agencies file and process their applications.[12] Sometimes, more than one bureaucratic procedure can be completed at the same time, thereby reducing the total time delay. In others cases, a precise procession of steps must be followed. Due to the fact that the time delay is often lengthy, some countries have implemented "single-window systems" to streamline the process of registering new firms. Around the world, there is significant variation in how long it takes to complete a procedural requirement to open a business. In 2012, the shortest time to complete a procedure was one day in New Zealand. The longest time was 694 days in Suriname.

I start by creating a *start-up regulation* measure that multiplies the number of procedures to open a business by the length of time (in days) to complete a procedure. Figure 2.3 displays the start-up regulation averages for four groups of countries: all oil producers in the developing world, oil producers in the developing world (except Gulf Arab oil states), nonproducers in the developing world, and OECD countries. A country is coded as an oil producer if it has at least 100 US dollars of oil and gas income per capita in a given year (see Table 2.1). As the bar plot reveals, oil producers have 60 percent higher start-up regulation on average than nonproducers. This value increases to 74 percent if we exclude the Gulf Arab oil states. The first row of Table 2.2 presents t-test results that confirm that the difference in means between oil producers and nonproducers in the developing world is statistically significant at a level of $p < 0.01$. Excluding the Gulf Arab oil states does not alter the level of statistical significance. Thus, an initial bit of evidence suggests that oil producers make it more difficult for domestic entrepreneurs to open a business.

Beyond start-up barriers, other aspects of business regulation affect the activities of domestic entrepreneurs and firms. Existing firms frequently need to acquire permits or licenses from the government, whereby they

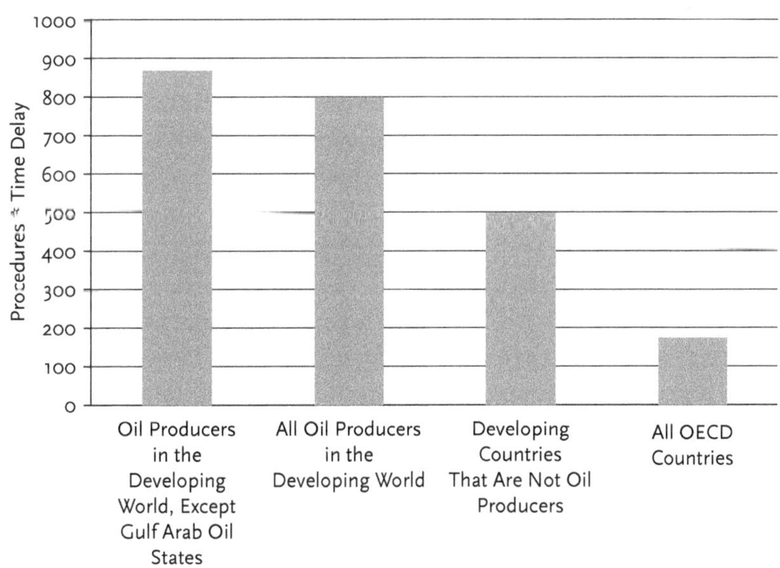

Figure 2.3 Business Start-up Regulation (averaged across groups)

Table 2.2 OIL AND BUSINESS REGULATION IN THE DEVELOPING WORLD

	Mean of all oil producers	Mean of oil producers (except Gulf Arab oil states)	Mean of nonproducers	Difference between all oil producers and nonproducers	Difference between oil producers (except Gulf Arab oil states) and nonproducers
Start-up Regulation	799.52	867.53	498.53	+ 300.99*** (60% higher)	+ 369*** (74% higher)
Permits Regulation	4,828.84	5,231.49	3,900.37	+ 928.47*** (24% higher)	+ 1,331.12*** (34% higher)
Tax Regulation	18,930.74	21,388.26	13,067.8	+ 5,862.94*** (45% higher)	+ 8,320.46*** (64% higher)
Contracts Regulation	28,388.22	28,430.08	24,466.85	+ 3,921.37*** (16% higher)	+ 3,963.23*** (16% higher)

*** significant at 0.01 level, in a one-tailed t-test.

are legally granted the permission to undertake a certain activity. I create a *permits regulation* measure, which multiplies the number of procedures and time delay for a small- or medium-sized business to obtain the necessary approvals to build and use a commercial warehouse.[13] The second row of Table 2.2 displays the differences between oil producers and nonproducers in permits regulation. On average, oil producers have 24 percent higher permits regulation than nonproducers. A t-test confirms that this difference is statistically significant at a $p < 0.01$ level.

Table 2.2 lists two other examples of business regulation that affect the activities of small- and medium-sized firms. *Tax regulation* multiplies the number of tax payments by the time (in estimated hours) to prepare documents, file, and pay taxes on profits, consumption, labor, and mandatory contributions.[14] This measure reflects the degree to which governments expect firms to comply with tax rules, and both component measures have a weakly positive correlation with the tax rates on business profits. On average, oil producers have 45 percent higher tax regulation than nonproducers. A t-test indicates that the difference in means between oil producers and nonproducers is statistically significant at a $p < 0.01$ level. I create another measure, *contracts regulation*, by multiplying the number of legal procedures by the time (in days) that it takes to legally resolve a commercial dispute.[15] A higher value suggests that resolving commercial disputes is more difficult. Oil producers have 16 percent higher contracts regulation than nonproducers. The t-test for the difference in means between oil producers and nonproducers is significant at a $p < 0.01$ level. Thus, the evidence compiled so far suggests that oil producers have more restrictive business environments for entrepreneurs and SMEs, specifically in terms of starting a business, obtaining permits, complying with tax procedures, and resolving contracts.

Over time, the imposition of higher regulatory barriers will have serious long-term negative consequences for private-sector development in oil producers. Although some entrepreneurs will simply endure higher start-up barriers and forge ahead, over time these barriers will discourage entrepreneurship among the population and decrease rates of new-firm entry. Researchers find that this effect holds even in industries with naturally high rates of entry.[16] Higher start-up regulation will also encourage entrepreneurs to open informal firms instead of formal ones, where they may be able to escape government bureaucracy but their legal protections are much weaker.[17] Research shows that countries with more tax procedures have lower rates of entrepreneurship, a link that holds irrespective of a country's tax rates.[18] Finally, the relationship between contract enforcement and broader developmental outcomes is no doubt complex,[19] but it is difficult to see how greater challenges in resolving commercial disputes will have a positive impact on the business community overall.

2.3 OIL AND CREDIT

Credit is often called the "lifeblood" of the private sector. Without access to credit, the vast majority of entrepreneurs are unable to open a business and existing firms face restrictions in their operations. In all countries, the government plays a critical role in credit lending by establishing the legal and information-based institutions that govern lending practices. Research shows that better legal protections enforced by the government have a positive relationship with the actual volume of lending.[20] In the developing world, many governments play a direct role in credit lending through state-owned banks. In 1970, state-owned banks in developing economies controlled an average of 67 percent of the banking system's total assets. Although this share declined to an average of 22 percent in 2009, in many places—such as China, India, Iran, and Saudi Arabia—the share of total assets controlled by fully or partially state-owned banks remains high.[21]

The developing world is routinely described as a place with a "credit crisis," and the lack of credit opportunities is what helped spark the microfinance movement that ramped up in the 1970s. While the long-term benefits of microlending continue to be debated by scholars,[22] the lack of credit and legal protections for borrowers has long been a barrier to entrepreneurial activities in the developing world.[23] Weak credit institutions also render credit markets more volatile in countries that are susceptible to economic shocks.[24] Developing countries have greater reason to fear economic shocks and oil producers are particularly vulnerable to shocks, given that their economies are closely linked to the price of a historically volatile commodity.[25]

I use two measures for exploring the differences in credit-lending patterns between oil producers and nonproducers. The first measure is *banking credit* and is an estimate of credit provided by the banking sector to various sectors (measured as a percentage of GDP). This measure does not include credit to the central government. The second measure is *private-sector credit* and is an estimate of the value of financial resources provided to the private sector through loans, purchases of nonequity securities, trade credits, and other accounts receivable that establish a claim for repayment (measured as a percentage of GDP). This measure may include credit to public enterprises. Credit-lending data are available for many developing countries starting in 1960, and overall the data set is 74 percent complete.[26]

Figure 2.4 plots *banking credit* for three groups in the developing world: long-term oil producers except the Gulf Arab oil states, the Gulf Arab oil states, and nonproducing countries. During the preboom era (before 1974) and the oil boom era (1974–1985), there are few observable differences in credit-lending patterns between non-Gulf Arab oil producers and nonproducers. The final years of the oil boom are even associated with increasing levels of credit from oil producers. But a divergence starts

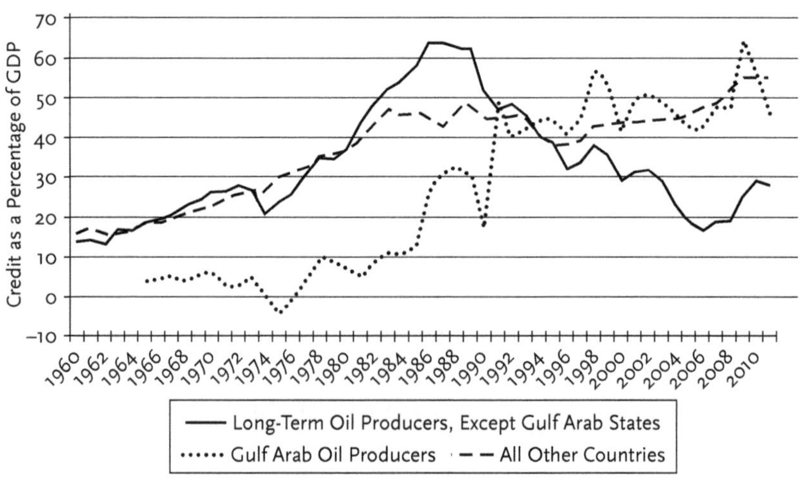

Figure 2.4 Domestic Credit by Banking Sector, Developing World Only, 1960–2011

to occur during the oil bust and afterward. On average, non-Gulf Arab oil producers provided rapidly declining levels of banking credit after 1988, with a slight recovery from 2009 to 2011. However, nonproducers provided steadily increasing credit over time. For the Gulf Arab oil states, the oil boom, oil bust, and beyond eras witnessed increasing levels of credit as a percentage of GDP. In fact, their lending after 1990 looks the same as non-producers, whereas all other oil producers venture down a different road.

Similar trends are observed in *private-sector credit* (see Figure 2.5), which is expected given that the two measures correlate at 0.75 from 1960 to 2011. Here, increased lending from Gulf Arab oil producers exceeds all other countries at an earlier time, by 1986, at the start of the oil bust. After 1991, three separate trajectories appear. Gulf Arab oil states led all developing countries in private-sector credit, while nonproducers provided higher levels of credit than all other oil producers. Private-sector credit among non-Gulf Arab oil producers reached its peak at the end of the oil boom era, but unlike the Gulf Arab oil states they have yet to surpass this level. In nonproducers, private-sector credit steadily increased after 1960.

Although Gulf Arab oil producers appear to be generous with their credit lending in recent years, oil producers as a whole have fostered weaker credit institutions than nonproducers. *Credit strength* is a measure that assesses the degree to which collateral and bankruptcy laws protect the rights of borrowers and lenders (scaled from 1 to 10, increasing values indicate greater protection).[27] The mean value of credit strength is 21 percent higher in nonproducers than oil producers. A t-test confirms that

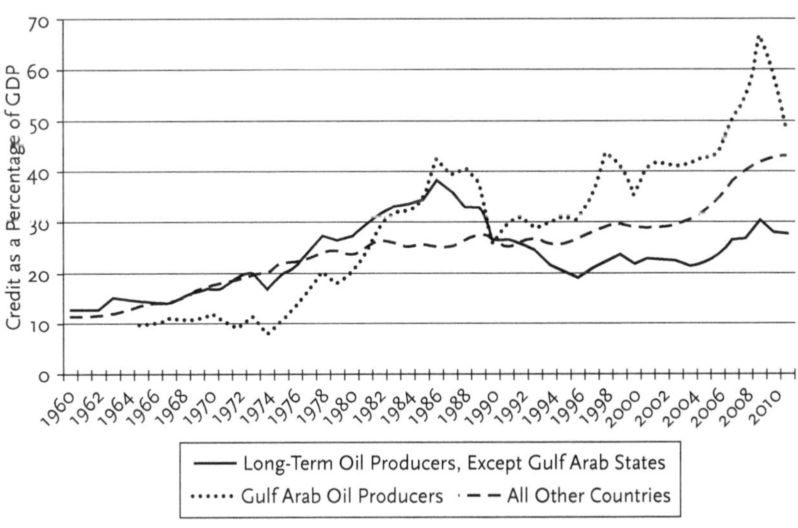

Figure 2.5 Domestic Credit to the Private Sector, Developing World Only, 1960–2011

this difference is statistically significant at the $p < 0.01$ level, holding even when the Gulf Arab oil states are omitted. *Depth of credit information* is a measure that evaluates the rules and practices affecting the coverage, scope, and accessibility of credit information from a public credit registry bureau (scaled from 1 to 6, increasing values indicate more information). Although the difference in means is not statistically significant, logistic regression demonstrates that being an oil producer is negatively associated with higher values.[28] That is, after controlling for other factors, oil producers are less likely than nonproducers to provide information about the rules and practices of credit lending.

2.4 STATISTICAL TESTS OF OIL'S RELATIONSHIP WITH THE BUSINESS ENVIRONMENT

The descriptive data and t-tests in the preceding text indicated that there are differences between developing countries that produce oil and those that do not. On average, oil producers impose more regulation on domestic entrepreneurs and SMEs in regard to firm start-up, acquiring permits, paying taxes, and resolving commercial disputes. Throughout the last two decades, credit lending among the non-Gulf Arab oil producers was considerably lower than credit lending among nonproducers or the Gulf Arab oil producers. As a whole, oil producers provide fewer legal protections for borrowers, and there is less information in public credit registry bureaus about the rules and practices affecting credit lending.

But other factors may be shaping these patterns. The presence of an authoritarian government, which is demonstrated to be the outcome of oil under certain conditions, could be influencing the observation of comparatively more restrictive business environments in oil producers. The level of income may influence these patterns given that wealthier developing countries should be more likely to adopt economic policies that are similar to OECD states, which have less restrictive business environments overall (refer to Figure 2.3). Political stability may matter because countries that experience endemic instability may have bigger concerns than reforming the business environment for domestic firms and entrepreneurs.

Using statistical techniques, I illustrate that the patterns outlined in the preceding text hold after controlling for these factors and other ones. Statistical techniques are additionally helpful in that they facilitate a closer understanding of the relationship between *oil income*—not simply being an oil producer—and the nature of the business environment. I start by employing methods for dealing with cross-sectional and panel data to investigate the link between oil income and the measures of regulation and credit introduced previously. I begin with estimating simple models with a small number of key independent variables.[29] I then add other control variables and employ a range of robustness checks to verify that the results are consistent when certain countries are removed, alternative measures are employed, and the data set is altered in other meaningful ways.

Along with a measure of *oil income*, I use two dummy variables to indicate if a country is an oil producer or a long-term oil producer. In the regressions, I always control for a country's *income per capita*, which is a measure of GDP (constant 2000-year dollars) divided by the total population and logged. We would expect wealthier developing countries to have lower regulatory and procedural barriers to entrepreneurs and SMEs, as well as to provide more credit to the private sector. Unless a per capita measure is used as the dependent variable, I control for the total population (logged) to account for the many small developing countries in the data set and the notion that reforms may be easier to implement in sparsely populated states (such as those in the Gulf Arab region).

To control for political factors, I include estimates of a country's *regime type* and *regime durable*. Higher values of regime type indicate a more democratic government (scaled from –10 to 10). Research finds that democracy is negatively associated with regulatory and procedural barriers to entrepreneurs.[30] Democracy should be associated with more credit lending. A squared regime type measure is used as a robustness check. *Regime durable* is a (logged) count of the number of years since the most recent regime change, or the end of a transition period that is defined by the lack of stable political institutions. Higher values should be associated with lower regulatory and procedural barriers as well as more credit to the private sector.[31]

Other control variables are included in the regressions, such as time dummy variables and period dummy variables that represent the oil boom and oil bust eras. Regional dummy variables denote to which of the six world regions a country belongs. Given the relevance of the Middle East/North Africa (MENA) region to any study of oil, I often substitute two MENA dummy variables: *Gulf Arab oil state* and *non-Gulf Arab MENA state*. I rerun statistical tests after removing all MENA countries from the data set to ensure that this region is not driving the results. Other controls are employed in the regressions on innovation, such as a country's literacy rate and per capita measures of power consumption. All statistical tests were conducted using Stata 12. See Table 2.3 for summary statistics.

Table 2.4 reports the findings when *start-up regulation* is the dependent variable.[32] I run an ordinary least squares (OLS) model that clusters by

Table 2.3 SUMMARY OF VARIABLES USED IN STATISTICAL TESTS

Variable	Observations	Mean	Standard deviation	Minimum	Maximum
Start-up Regulation[a]	1088	5.553	1.265	0	9.107
Permits Regulation[a]	798	7.969	0.744	5.765	10.575
Tax Regulation[a]	797	8.811	1.198	4.682	12.633
Contracts Regulation[a]	1083	9.929	0.558	7.832	11.427
Income[a]	7294	8.293	1.172	5.426	11.854
Population[a]	8715	15.623	1.71	10.715	21.014
Regime Type	7275	0.526	7.486	-10	10
Regime Durable[a]	7344	2.435	1.293	0	5.308
Banking Credit	6292	73.14	517.34	-72.99	17521.27
Private-Sector Credit	6301	56.902	440.017	0.557	15788.26
Management Time	247	8.526	6.323	0	33.6
Informal Payments	243	33.714	22.761	0	98.26
New-Firm Density	689	2.886	4.281	0.002	38.999
Patent Applications[a]	3556	5.484	2.747	0	12.859
Literacy	3420	81.274	20.816	19.032	100
Long-Term Oil Producer	8721	0.199	0.399	0	1
Oil Producer	8721	0.222	0.415	0	1
Oil Income[a]	7896	2.36	2.897	0	11.147
Establishment Permits (Iran)	50	8027.4	12418.86	23	54288
Operation Permits (Iran)	50	1693.02	2183.613	28	8135
Oil Income (Iran)	47	900.681	782.880	237.626	3325.12
Regime Type (Iran)	47	-6.021	4.489	-10	3
Population (Iran)[a]	50	17.582	0.39	16.907	18.108
Change in GDP pc (Iran)	46	112.413	576.394	-1491.15	2268.709

Note: A superscript of "a" indicates the variable was logged before the analysis.

country and includes region dummy variables, the full set of time dummy variables, robust standard errors, and one-period lags on all independent variables.[33] I start with a basic model that includes only *oil income* as an independent variable (column 1), which emerges as positive and statistically significant at a $p < 0.01$ level. This suggests that more oil is associated with higher start-up regulation. *Oil income* retains its statistical significance when the other key controls (*income, population, regime type*, and *regime durable*) are included in the model (columns 2, 3, and 4). Using a squared measure of *regime type* does not alter the results, nor does substituting *oil producer* as the measure of oil wealth. Removing all MENA countries (column 5) slightly increases the magnitude of the *oil*

Table 2.4 OIL AND START-UP REGULATION IN THE DEVELOPING WORLD

Independent variable	*Start-up regulation* (1)	*Start-up regulation* (2)	*Start-up regulation* (3)	*Start-up regulation* (4)	*Start-up regulation* (5)	*Start-up regulation* (6)
Oil Income (log)	0.107***	0.122***	0.116***	0.116***	0.117***	0.119***
	(0.025)	(0.028)	(0.031)	(0.031)	(0.034)	(0.031)
Income (log)	—	−0.247**	−0.244**	−0.246**	−0.214	−0.239 *
	—	(0.112)	(0.117)	(0.115)	(0.136)	(0.122)
Population (log)	—	−0.026	−0.027	−0.027	−0.034	−0.031
	—	(0.05)	(0.051)	(0.052)	(0.057)	(0.054)
Regime Type	—	—	−0.019	−0.019	−0.023	−0.02
	—	—	(0.014)	(0.015)	(0.015)	(0.015)
Regime Durable	—	—	—	0.004	−0.001	0.008
(log)	—	—	—	(0.058)	(0.066)	(0.06)
Gulf Arab Oil State	—	—	—	—	—	-1.766***
	—	—	—	—	—	(0.46)
Non-Gulf Arab	—	—	—	—	—	-1.577***
MENA State	—	—	—	—	—	(0.276)
Constant	6.438***	8.965***	9.109***	9.117***	8.963***	9.116***
	(0.194)	(1.32)	(1.35)	(1.349)	(1.466)	(1.353)
Years	2004–2010	2004–2010	2004–2010	2004–2010	2004–2010	2004–2010
Countries	124	124	120	120	105	120
Regions	Non-OECD	Non-OECD	Non-OECD	Non-OECD	Non-OECD, no MENA	Non-OECD
Observations	852	816	789	789	693	789
R-squared	0.26	0.288	0.328	0.328	0.294	0.329

The results are from an OLS regression with robust standard errors, country clusters, year dummies, and regional dummies (not reported). All independent variables were lagged one period. Non-OECD countries only. * sig at 0.1; ** sig at 0.05; *** sig at 0.01.

income coefficient, as does splitting the MENA region dummy into two groups. In fact, the MENA region is associated with lower levels of start-up regulation.

Thus, the statistical tests help confirm that oil wealth is positively associated with the level of government regulation of domestic entrepreneurs seeking to start a business, after controlling for other key factors.[34] Additionally, I show that this relationship between oil and business regulation is not simply another manifestation of authoritarianism or a phenomenon that is specific to countries located in the Middle East.

I use the same approach and estimating model to examine the relationship between oil and *permits regulation, tax regulation*, and *contracts regulation* (see Table 2.5). In the case of *permits regulation*, the findings indicate a positive association but they are somewhat more attenuated. *Oil income* is significant at a level of $p < 0.03$ when it is the only independent variable in the regression equation. However, when the other economic, political, and demographic controls are included, its level of significance weakens to $p < 0.06$. Removing all MENA countries does not change the results, nor does substituting a squared measure of *regime type*. The Gulf Arab oil states are negatively associated with permits regulation at a $p < 0.04$ level. When *oil producer* is substituted, the variable is positive but statistically insignificant. These findings provide some evidence that oil income is positively associated with the level of regulation of SMEs seeking a permit to build a warehouse. Simply being an oil producer, however, is not associated with the level of permits regulation. Although t-tests indicated a statistically significant difference of means, the link between oil and permits regulation is weaker under the scrutiny of regression analysis. That said, the findings do confirm that Gulf Arab oil states, in particular, are associated with lower levels of permits regulation.

The preliminary t-test results are strongly confirmed in the cases of *tax regulation* and *contracts regulation*. As columns 5 and 8 display, simply being an oil producer is positively associated with the level of tax and contracts regulation at the $p < 0.05$ and $p < 0.02$ levels of statistical significance. Using *oil income* for the *tax regulation* regression (column 4) achieves a $p < 0.02$ level of significance, removing all MENA countries slightly increases the magnitude of the coefficient, and substituting a squared regime type measure does not change the results. Although the Gulf Arab oil states are negatively associated with tax regulation, there appears to be some evidence of a positive association with contracts regulation.

Overall, there is consistent evidence of a positive relationship between oil wealth and start-up regulation, tax regulation, and contracts regulation in the developing world. As an additional robustness check, I use the mean values of all independent and dependent variables and perform cross-sectional regressions using region dummies and robust standard errors. The results are consistent with those previously observed (see Table 2.6).

To investigate the relationship between oil and credit, I use panel data methods given that I have data on some countries that reach back to 1960. I run a generalized least squares (GLS) model using a Prais-Winsten transformation for an AR(1) process and include a lagged dependent variable, region dummy variables, a full set of time (or period) dummy variables, robust standard errors, and one-period lags on all independent variables.

Table 2.5 OIL AND REGULATION OF PERMITS, BUSINESS TAXATION, AND CONTRACT ENFORCEMENT

Independent variable	*Permits regulation* (1)	*Permits regulation* (2)	*Permits regulation* (3)	*Tax regulation* (4)	*Tax regulation* (5)	*Tax regulation* (6)	*Contracts regulation* (7)	*Contracts regulation* (8)	*Contracts regulation* (9)
Income (log)	−0.058	−0.031	−0.131	−0.362***	−0.326***	−0.407***	−0.153*	−0.123	−0.164*
	(0.084)	(0.084)	(0.085)	(0.123)	(0.125)	(0.134)	(0.088)	(0.083)	(0.093)
Population (log)	0.083 *	0.097*	0.068	0.126***	0.144***	0.109**	−0.002	0.015	−0.005
	(0.043)	(0.044)	(0.044)	(0.046)	(0.047)	(0.047)	(0.034)	(0.034)	(0.035)
Regime Type	0.01	0.008	0.008	−0.012	−0.015	−0.015	0.026***	0.023***	0.027***
	(0.01)	(0.01)	(0.01)	(0.017)	(0.018)	(0.018)	(0.009)	(0.009)	(0.01)
Regime Durable (log)	−0.074	−0.078	-0.011	0.133**	0.129 *	0.176**	0.022	0.016	0.02
	(0.059)	(0.058)	(0.063)	(0.063)	(0.065)	(0.07)	(0.036)	(0.036)	(0.04)
Oil Income (log)	0.038 *	—	0.04 *	0.075**	—	0.08**	0.056***	—	0.06***
	(0.02)	—	(0.021)	(0.031)	—	(0.034)	(0.02)	—	(0.022)
Oil Producer	—	0.129	—	—	0.376**	—	—	0.244**	—
	—	(0.13)	—	—	(0.185)	—	—	(0.1)	—
Gulf Arab Oil State	−0.657**	−0.549*	—	-3.509***	-3.319***	—	0.275	0.421*	—
	(0.308)	(0.314)	—	(0.498)	(0.499)	—	(0.213)	(0.226)	—
Non-Gulf Arab MENA State	−0.263	−0.262	—	−0.974***	−0.975***	—	0.117	0.117	—
	(0.254)	(0.252)	—	(0.301)	(0.3)	—	(0.167)	(0.163)	—
Constant	7.439***	7.08***	8.141***	10.302***	9.799***	10.84***	11.147***	10.705***	11.268***
	(1.11)	(1.1)	(1.115)	(1.261)	(1.208)	(1.33)	(0.917)	(0.836)	(0.962)
Years	2006–2010	2006–2010	2006–2010	2006–2010	2006–2010	2006–2010	2004–2010	2004–2010	2004–2010
Countries	118	118	103	119	119	104	120	120	105
Regions	Non-OECD	Non-OECD	Non-OECD, no MENA	Non-OECD	Non-OECD	Non-OECD, no MENA	Non-OECD	Non-OECD	Non-OECD, no MENA
Observations	576	576	505	580	580	509	785	785	690
R-squared	0.232	0.22	0.186	0.474	0.467	0.244	0.358	0.335	0.358

The results are from an OLS regression with robust standard errors, country clusters, time dummies, region dummies, and one-period lags on all independent variables. * sig at 0.1; ** sig at 0.05; *** sig at 0.01.

Table 2.6 OIL AND BUSINESS REGULATION, CROSS-SECTIONAL REGRESSIONS USING MEAN VALUES

Independent variable	*Start-up regulation* (1)	*Permits regulation* (2)	*Tax regulation* (3)	*Contracts regulation* (4)
Income (log)	-0.296**	-0.036	-0.37***	-0.163*
	(0.115)	(0.088)	(0.122)	(0.0878)
Population (log)	-0.028	0.084**	0.126***	-0.0238
	(0.058)	(0.042)	(0.046)	(0.035)
Regime Type	-0.02	0.009	-0.014	0.03***
	(0.017)	(0.011)	(0.019)	(0.01)
Regime Durable (log)	0.014	-0.086	0.169**	0.052
	(0.077)	(0.073)	(0.084)	(0.049)
Oil Income (log)	0.123***	0.036*	0.069**	0.062***
	(0.032)	(0.02)	(0.03)	(0.02)
Gulf Arab Oil State	-1.925***	-0.826**	-3.557***	0.242
	(0.57)	(0.337)	(0.52)	(0.225)
Non-Gulf Arab MENA State	-1.29***	-0.381	-1.003***	0.128
	(0.315)	(0.258)	(0.317)	(0.172)
Constant	9.186***	7.263***	10.146***	11.476***
	(1.358)	(1.089)	(1.193)	(0.942)
Years	2004–2010	2006–2010	2006–2010	2004–2010
Observations	120	119	120	120
R-squared	0.304	0.252	0.513	0.376

The results are from an OLS regression with robust standard errors. Other regional dummies were included (not reported). Non-OECD countries only. * sig at 0.1; ** sig at 0.05; *** sig at 0.01.

Table 2.7 reports the findings for *banking credit* and *private-sector credit*. Both *oil income* and *oil producer* emerge as negative and statistically significant predictors of credit lending at a $p < 0.01$ level. These associations hold after controlling for *regime type, regime durable, income, population,* and all regional, temporal, and oil boom/bust variables. Removing the MENA region decreases the magnitude of the oil coefficient, but it does not alter its level of statistical significance. Adding the oil boom and oil bust variables generates contrasting relationships. Whereas the oil boom is associated with higher levels of banking credit, it is associated with lower private-sector credit. However, the oil bust is associated with lower levels of both credit measures.

Given the contours of Figures 2.4 and 2.5, it is useful to examine changes in the *oil producer* coefficient during different time periods of interest. Looking at *private-sector credit* and truncating the period of analysis to only the preboom era reveals that the *oil producer* coefficient is positive but

Table 2.7 OIL AND CREDIT IN THE DEVELOPING WORLD

Independent variable	*Banking credit* (1)	*Banking credit* (2)	*Banking credit* (3)	*Banking credit* (4)	*Private-sector credit* (5)	*Private-sector credit* (6)	*Private-sector credit* (7)	*Private-sector credit* (8)
Income (log)	1.24***	1.441 ***	1.29***	0.983***	0.658***	0.681***	0.662***	0.558***
	(0.275)	(0.333)	(0.278)	(0.248)	(0.151)	(0.186)	(0.15)	(0.129)
Population (log)	0.266 **	0.25 *	0.284 **	0.174	0.05	0.07	0.058	0.018
	(0.132)	(0.136)	(0.135)	(0.128)	(0.069)	(0.073)	(0.069)	(0.072)
Regime Type	−0.053**	-0.063 **	-0.033	−0.039	−0.003	−0.005	0.011	0.002
	(0.027)	(0.026)	(0.026)	(0.026)	(0.014)	(0.015)	(0.013)	(0.014)
Regime Durable (log)	0.183	0.2[illegible]4	0.261 *	0.217	0.086	0.063	0.119	0.096
	(0.149)	(0.135)	(0.148)	(0.148)	(0.075)	(0.081)	(0.074)	(0.075)
Oil Income (log)	−0.357***	−0.286***	−0.324***	—	-0.115***	-0.102 **	−0.087 **	—
	(0.074)	(0.075)	(0.073)	—	(0.043)	(0.045)	(0.041)	—
Oil Producer	—	—	—	-1.775 ***	—	—	—	−0.528**
	—	—	—	(0.409)	—	—	—	(0.243)
Oil Boom Era	—	—	0.871***	—	—	—	−0.339**	—
	—	—	(0.288)	—	—	—	(0.162)	—
Oil Bust Era	—	—	−0.925***	—	—	—	−0.933***	—
	—	—	(0.321)	—	—	—	(0.189)	—
Constant	-11.807***	-13.482***	-12.577***	-8.613**	-5.609***	−6.083***	-5.334***	−4.446***
	(3.839)	(4.231)	(3.87)	(3.618)	(1.722)	(1.922)	(1.71)	(1.676)
Years	1960–2010	1960–2010	1960–2010	1960–2010	1960–2010	1960–2010	1960–2010	1960–2010
Countries	123	106	123	123	123	106	123	123
Regions	Non-OECD	Non-OECD, no MENA	Non-OECD	Non-OECD	Non-OECD	Non-OECD, no MENA	Non-OECD	Non-OECD
Observations	4306	37[illegible]2	4306	4307	4310	3718	4310	4133
Durbin-Watson	1.881	1.897	1.885	1.88	1.914	1.924	1.911	1.914

The results are from an AR(1) Prais-Winsten regression with robust standard errors, a one-period lagged dependent variable (not reported), independent variables lagged one period, and year and regional dummies (both not reported and the former omitted when period dummies are used). * sig at 0.1; ** sig at 0.05; *** sig at 0.01.

statistically insignificant. During the oil boom and the postbust years, the *oil producer* coefficient is again positive but insignificant. During the oil bust, the *oil producer* coefficient is negative and highly statistically significant. Similar findings occur with respect to *banking credit* and when substituting *oil income*. Thus, it seems that the negative association of oil with credit is largely driven by the highly negative association during the oil bust era. This indicates that oil producers restricted credit lending when times were bad and oil prices were low. But when times were good, such as prior to the oil boom and following the oil bust, credit lending in oil producers was indistinguishable from nonproducers. Most surprisingly, when times were *exceptionally good* during the oil boom (what Terry Lynn Karl described as "the largest transfer of wealth ever to occur without war"[35]), credit lending in oil producers was indistinguishable from credit lending in nonproducers.

2.5 MANAGERIAL BURDEN AND BRIBERY BY FIRMS

More government regulation in the areas of permits, taxes, contract resolution, and credit access will affect the day-to-day activities of existing firms and their managers. If the government requires that firms complete more regulatory procedures, file more documents, and spend more time waiting for the wheels of bureaucracy to turn, there should be an increase in the time that firm managers report having to spend on these activities. We would also expect to see firm owners resorting to bribery of public officials to get around various regulations. Although cross-national data for such activities are limited, an analysis of the available data indicates a positive association between oil and the amount of time that firm managers report having to deal with government regulations. In addition, there is a positive association between oil and the percentage of firm owners who report bribing officials in order to bypass regulation.

To demonstrate this, I utilize the World Bank's *Enterprise Surveys* data set, which includes an estimate of how much time in a week firm managers report having to spend on the requirements imposed by government regulations (i.e., taxes, customs, labor regulations, licensing, and registration). Among developing countries, this measure of *managerial burden* varies from 0 percent of management time to as high as 33.6 percent. The data set contains estimates for ninety-nine developing countries taken at some point between 2002 and 2010. Among this group, thirty-four countries are classified as oil producers during at least one of these years. Sixteen countries are long-term oil producers. The MENA oil producers in the sample population are Algeria, Egypt, Syria, and Yemen.

The mean value of *managerial burden* is 10.11 percent for oil producers and 8.35 percent for nonproducers. This difference is statistically significant at a $p < 0.04$ level, suggesting that managers of firms that operate in oil-producing countries devote more time per week dealing with government regulation on average. The difference is larger between long-term oil producers and all other developing countries. The mean value for long-term oil producers is 10.68 percent as opposed to 8.51 percent for all other developing countries, a difference that is statistically significant at a level of $p < 0.01$. Indeed, running a business that just happens to be located in an oil-producing country is associated with spending more time dealing with regulation, a finding that is entirely consistent with the results outlined in the preceding text.

To fully investigate the relationship between oil and *managerial burden*, I run a multilevel model regression using maximum likelihood estimation. Multilevel modeling is a preferred method over the clustered standard errors model when a data set is highly unbalanced.[36] Regions and countries are specified as the levels of analysis. I include the core set of controls, center the time variable, and lag the independent variables by one period. The results listed in Table 2.8 show that *oil income* is positively associated with *managerial burden* at a $p < 0.09$ level of statistical significance (column 1). This suggests a slightly more attenuated association. However, when *long-term oil producer* is substituted in column 3, the coefficient emerges as positive and statistically significant at a level of $p < 0.03$. All else equal, countries that are long-term oil producers place greater managerial burdens on firms.

But managers have ways of dealing with regulation. In any country, rich or poor, democracy or autocracy, some managers will consider bribing public officials as a way to get around regulation. To explore the relationship between oil and bribery by firms, I use the *Enterprise Survey*'s measure of the percentage of firms that report being expected to make informal payments to public officials to "get things done" with regard to customs, taxes, licenses, regulations, services, and so forth. Among developing countries, this measure of *bribery* ranges from 0 percent to 98 percent. There are estimates for 108 developing countries, taken at various times between 2002 and 2010. The sample population includes thirty-five oil producers, of which seventeen are long-term oil producers. The MENA oil producers in the data set are Algeria, Egypt, Oman, Syria, and Yemen.

The t-tests generate mixed results about the link between oil and bribery. A t-test confirms no statistically significant difference between the means of oil producers (35.17) and nonproducers (35.9). However, just as in the case of *managerial burden* examined previously, a t-test confirms a difference at a level of $p < 0.04$ in the means of long-term oil producers (42.48) and all other developing countries (34.36). I utilize multilevel

Table 2.8 OIL, MANAGEMENT TIME DEALING WITH PUBLIC OFFICIALS, AND BRIBES TO PUBLIC OFFICIALS

	Independent variable	*Mgmt time* (1)	*Mgmt time* (2)	*Mgmt time* (3)	*Bribes* (4)	*Bribes* (5)	*Bribes* (6)
Fixed	Constant	8.041	3.825	5.467	116.396***	95.807***	101.427***
Effects		(8.319)	(7.743)	(7.413)	(25.51)	(23.578)	(22.971)
Country	Income (log)	-1.12*	-0.862	-1.016*	-9.003***	-7.597***	-8.281***
Level		(0.638)	(0.614)	(0.6)	(1.871)	(1.749)	(1.754)
	Population (log)	0.535	0.705**	0.662**	-0.158	0.629	0.533
		(0.351)	(0.329)	(0.312)	(1.167)	(1.113)	(1.074)
	Oil Income (log)	0.343*	—	—	1.411**	—	—
		(0.201)	—	—	(0.686)	—	—
	Oil Producer	—	1.174	—	—	2.487	—
		—	(1.064)	—	—	(3.499)	—
	Long-Term Oil Producer	—	—	2.947**	—	—	8.659 *
		—	—	(1.272)	—	—	(4.608)
	Regime Type	0.162*	0.139	0.18**	-0.917***	-1.067***	-0.935***
		(0.086)	(0.086)	(0.086)	(0.29)	(0.286)	(0.289)
	Regime Durable (log)	-0.215	-0.25	-0.224	-1.445	-1.629	-1.444
		(0.46)	(0.463)	(0.456)	(1.413)	(1.423)	(1.415)
Random	Variance Country	2.105	2.278	2.128	13.03	13.242	13.007
Effects		(0.805)	(0.746)	(0.762)	(1.537)	(1.558)	(4.095)
	Variance Region	1.966	1.99	1.921	0.000	0.000	0.000
		(0.753)	(0.765)	(0.742)	(0.000)	(0.000)	(0.000)
	Correlation Country	0.139	0.161	0.144	0.514	0.518	0.51
	Correlation Region	0.124	0.128	0.12	0.000	0.000	0.000
	Countries	114	114	114	115	115	115
	Observations	232	232	232	228	228	228

The results are from a multilevel (hierarchical) model using maximum likelihood estimation that defines region and country as the levels of analysis. The model includes a variable that centers the year (not reported). Logging the dependent variables does not change the results. Time period is 2002–2010. * sig at 0.1; ** sig at 0.05; *** sig at 0.01.

modeling techniques to investigate the nature of this link and to control for other factors. Table 2.8 reports that *oil income* emerges as positive and statistically significant at a $p < 0.04$ level of significance (column 4). When *long-term oil producer* is substituted, it is positive and statistically significant at a level of $p < 0.06$. Thus, I confirm that the level of oil income (and perhaps simply being a long-term oil producer) is linked to a higher percentage of firms that report bribing officials to "get things done" with respect to regulation.

It should be noted that, while bribery seems like a practical solution for firms that are overburdened by regulation, studies show that bribery imposes serious long-term costs on firms. A study of Ugandan firms demonstrated that a one percentage point increase in the rate of bribery is associated with a three percentage point decrease in firm growth. This effect is estimated to be three times that of taxation.[37] Bribery may solve short-term problems with regulation but it can impose long-term costs.

2.6 OIL, ENTREPRENEURSHIP, AND INNOVATION

Higher regulatory barriers, less access to credit, and the expectation of bribery should have a negative effect on entrepreneurship rates in oil-producing countries. Yet, many budding entrepreneurs will simply endure a more restrictive business environment and push ahead toward start-up anyway. Some may choose to seek out informal credit markets or rely on family and friends for loans, as many entrepreneurs do. Some will bribe government officials to get around regulation. However, as this decision-making process plays out across a population over time, we should see comparatively fewer entrepreneurs emerging in countries that impose higher regulatory barriers, such as oil producers. Using recently released data on new-firm entry rates, I confirm that this is precisely the case.

I rely on estimates of the number of newly registered limited-liability firms in a country as a ratio of the working-age population (per one thousand people aged fifteen to sixty-four).[38] At least two years of data are available for seventy-six developing countries starting in 2004. The MENA oil producers that are represented in the sample population are Algeria, Egypt, Iraq, Oman, Syria, Tunisia, and the United Arab Emirates. In a preliminary analysis of the data set conducted by researchers at the World Bank, a negative relationship was observed between regulatory barriers and new-firm entry rates.[39] This implies that if oil is linked to higher regulatory barriers, oil should also be linked to lower new-firm entry rates.

I begin with simple t-tests to examine the relationship between oil and new-firm entry rates in the developing world. Whereas the mean for oil producers is 1.42 new firms per one thousand working-age individuals, nonproducers have a mean that is substantially higher at 2.68 new firms per one thousand working-age individuals. A t-test confirms that this difference is statistically significant at a level of $p < 0.01$. The findings are suggestive, but somewhat less robust, for long-term oil producers. Long-term oil producers have a mean of 1.66 new firms per one thousand working-age individuals and all other developing countries have a mean of 2.36, a difference that is statistically significant at a level of $p < 0.08$.

Given the size of the data set and its balanced nature, I employ an OLS regression model that clusters by country and is estimated with region dummy variables, a full set of year dummy variables, one-period lags on all independent variables, and robust standard errors.[40] The results are reported in Table 2.9, columns 1 through 3. *Oil income* emerges as a negative and statistically significant predictor of new-firm entry rates ($p < 0.02$) after controlling for other factors. Using a squared measure of regime type or logging the dependent variable does not alter the results. Splitting the MENA region into the Gulf Arab oil states and all other MENA countries (column 2) suggests that the Gulf Arab oil states are associated with lower new-firm entry rates (the level of significance is $p < 0.07$). This finding holds when *oil producer* is substituted in column 3, which is significant at a $p < 0.04$ level. Removing all MENA countries does not affect the statistical significance of *oil income* or *oil producer*, and the magnitude of the coefficient is about 20 percent higher.

These tests reveal that oil income—or simply being an oil producer—is negatively associated with a country's new-firm entry rate. Although Gulf Arab oil states were observed to have lower levels of business regulation and higher levels of credit lending, they are associated with lower rates of new-firm entry. This may be a reflection of the fact that the reforms adopted by some of the Gulf Arab oil states are recent and have yet to impact rates of new-firm entry. More data collection over time will help to corroborate or challenge these findings, but it is important to note how they contribute to an explanation of the patterns discussed at the beginning of the chapter. The slow convergence of nonoil incomes in oil producers and nonproducers throughout the last half century is partly explained due to the fact that oil producers have charted lower rates of entrepreneurship.

If oil producers have more restrictive business environments, provide less credit, and have lower rates of entrepreneurship, we should witness less innovation in these economies over time. Indeed, many new firms open because they are in sectors at the edge of the technological frontier where market demand is expanding, such as biomedicine, computer software, or information and communication technologies. Some new firms open because they

Table 2.9 OIL, ENTREPRENEURSHIP, AND INNOVATION IN THE DEVELOPING WORLD

Independent variable	*New firms (1)*	*New firms (2)*	*New firms (3)*	*Patent applications (4)*	*Patent applications (5)*	*Patent applications (6)*
Income (log)	2.305***	2.341***	2.112***	1.051***	1.108***	1.029***
	(0.565)	(0.577)	(0.556)	(0.236)	(0.238)	(0.234)
Regime Type	0.113*	0.109*	0.133*	0.016	0.013	0.017
	(0.063)	(0.063)	(0.073)	(0.015)	(0.015)	(0.016)
Regime Durable (log)	0.675*	0.686*	0.859**	0.048	0.058	0.076
	(0.354)	(0.355)	(0.41)	(0.067)	(0.066)	(0.07)
Population (log)	—	—	—	1.331***	1.319***	1.244***
	—	—	—	(0.112)	(0.111)	(0.102)
Literacy	—	—	—	0.039***	0.038***	0.037***
	—	—	—	(0.01)	(0.01)	(0.01)
Oil Income (log)	-0.457**	-0.449**	—	-0.153***	-0.144***	—
	(0.175)	(0.176)	—	(0.048)	(0.048)	—
Oil Producer	—	—	-1.578**	—	—	-0.528**
	—	—	(0.74)	—	—	(0.215)
MENA Region	-1.168	—	—	-1.361***	—	—
	(0.746)	—	—	(0.342)	—	—
Gulf Arab Oil States	—	-1.808*	-3.182***	—	-2.766***	-3.109***
	—	(0.948)	(1.178)	—	(0.478)	(0.503)
Non-Gulf Arab MENA State	—	-1.03	-1.203	—	-1.304***	-1.383***
	—	(0.772)	(0.752)	—	(0.34)	(0.346)
Constant	-17.579***	-17.939***	-17.251***	-27.199***	-27.338***	-25.406***
	(4.28)	(4.425)	(4.529)	(3.057)	(3.027)	(2.759)
Years	2005–2010	2005–2010	2005–2010	1963–2010	1963–2010	1963–2010
Countries	70	74	70	97	97	97
Observations	443	443	443	1978	1978	1978
R squared	0.423	0.424	0.386	0.729	0.733	0.726

The results are from an OLS regression with robust standard errors, country clusters, independent variables lagged one period, year dummies, and regional dummies (not all reported). Non-OECD countries only. * sig at 0.1; ** sig at 0.05; *** sig at 0.01.

utilize the latest and most innovative technology to provide a particular good or service. Some open as the direct result of an invention. The fact that many of these innovative firms begin as SMEs is also important.

There is evidence that a country's natural resource endowment affects innovation,[41] but this relationship has not been systematically tested using multiple regression on a large sample of developing countries. In order to provide the first such test, I use a measure that has the broadest coverage across the developing world and over time: the number of patent applications by residents of a country.[42] Although not all patent applications are ultimately granted, entrepreneurs often raise capital on the basis of patent applications and enter into licensing agreements prior to the actual awarding of a patent.[43] I assemble a data set of ninety-seven developing countries over the period from 1963 to 2010 that includes twenty-two long-term oil producers.[44] I log transform the patents measure and divide it by the country's population to create a measure of *per capita patent applications by residents*.[45] Nonproducers, as a group, have a 100 percent higher value of *per capita patent applications by residents* than oil producers. The t-test results indicate that the difference in means between nonproducers and oil producers is statistically significant at a $p < 0.01$ level.[46] In the developing world, it seems that oil producers generate fewer patent applications than nonproducers.

I next turn to regression techniques to better evaluate this relationship. I run an OLS model that clusters by country and includes region dummy variables, the full set of year dummy variables, robust standard errors, and one-period lags on all independent variables. I also control for the education level in a country, given the large body of research on how education affects innovation.[47] I use the most widely available measure of education: the percentage of the adult population that is literate.[48] In the robustness checks, I control for a country's total consumption of electric power (measured in kWh), given that this is a necessary component of technological activities.[49]

The findings reveal that *oil income* is negatively associated with patent applications by residents at a $p < 0.01$ level of statistical significance (columns 4–6 of Table 2.9).[50] The MENA region is associated with fewer patent applications. Removing all MENA countries reduces the magnitude of *oil income* by about 20 percent, but it is still negative and statistically significant at a level of $p < 0.03$. Using the squared measure of regime type does not change the results. The results do not change when a GLS model is estimated using a Prais-Winsten transformation or a multilevel model is employed using maximum likelihood estimation.

Overall, the findings are highly supportive of the notion that oil is associated with fewer patent applications by residents. This is consistent with the earlier finding that rates of entrepreneurship are lower in oil producers.

Indeed, given fewer new firms, we should correspondingly witness fewer demonstrations of innovation such as patent applications. Fewer patent applications may also be a direct outcome of a more restrictive business environment. If owners and managers are forced to devote more time to dealing with regulation and overcoming barriers to performing their daily activities, they will have fewer resources or capacity to engage in innovation. Lower rates of innovation sustained over time will threaten the ability of oil producers to diversify their economies.

2.7 CONCLUSION

The beginning of this chapter set the stage by documenting the gradual stagnation of nonoil incomes in oil producers throughout the last half century. Given the vast riches these countries reaped during this period, it may come as quite a shock to see them experiencing extreme difficulties in expanding the nonoil portions of their economies compared to oil-poor countries—especially after the oil boom of the 1970s. To date, the standard explanation for this trend is the Dutch Disease phenomenon, mismanaged oil revenues, and inefficient state-owned enterprises, or it is chalked up to the rentier state effect. This chapter's findings provide us with another way to understand this trend. Oil is strongly associated with a wide range of government-imposed barriers on the activities of domestic entrepreneurs and SMEs. Managers spend more time dealing with government regulation and they are more likely to bribe officials if their firm happens to be located in an oil-producing country. It then comes as little surprise to see oil producers charting lower rates of entrepreneurship and filing fewer patent applications. If oil producers do not take appropriate measures to counteract these trends, they will only lead to a prolonged stagnation of their nonoil incomes into the future.

Chapter 3 outlines a theory that seeks to explain this association between oil and a restrictive business environment. I argue that an oil windfall dampens the *interests* of policy makers and business elites about the supply of business environment reforms and blunts the *capacity* of nonelites to demand reforms. I then test my theory through a series of case studies, each one designed to address a specific aspect of the link between an oil windfall and the adoption of government policies that define the business environment. Case-study analysis proves to be a powerful tool for evaluating the microfoundations of my theory and getting us to a clearer understanding of why, and to what extent, oil producers place restrictions on the activities of domestic entrepreneurs and SMEs.

CHAPTER 3

A Theory about Oil, Policy Makers, and the Business Environment

Chapter 2 demonstrated that oil is associated with six outcomes that are linked to broader negative trends in economies: the imposition of more regulatory restrictions on entrepreneurs and SMEs, less credit lending as a percentage of GDP, more time spent by managers dealing with regulation, more bribery of public officials by firms, lower rates of entrepreneurship, and fewer innovative activities. These associations hold after controlling for confounding variables, subjecting the results to robustness checks, and using alternative measures of oil wealth. The analysis confirmed that the Gulf Arab oil states are exceptions to these trends, but only in terms of regulatory policies and credit lending over the last two decades.

This chapter explains *why* this is happening. I argued in Chapter 1 that existing theories provide us with limited insights and conflicting predictions about how oil producers make policy decisions affecting the private sector. Although it is common to characterize oil producers as countries with irrational and incoherent economic policies, I argue that a logic underpins their decision making, and in particular the decisions that impact the private sector. Returning to my original question: If oil producers have charted unimpressive economic performance and witnessed stagnation in their nonoil sectors throughout the last half century, why do their governments continue to make things harder by excessively regulating entrepreneurs and SMEs? I offer a probabilistic theory to answer this question, focusing on how an oil windfall sets in motion causal mechanisms that dampen the *interests* of policy makers and elites about supplying reforms and blunt the *capacity* of nonelites to demand these reforms. Understanding these causal mechanisms helps to explain the trends outlined in the preceding text, and assists in the task of thinking about ways to minimize the broader challenges that confront oil producers today.

My theory contains a "supply side" component, which holds that an oil windfall affects the *interests* of both policy makers and business elites with respect to business environment reforms. I argue that an oil windfall leads policy makers to develop interests that are opposed to reforms that lower the barriers to market entry, improve access to credit, streamline the permit application process, and more generally ease government-imposed restrictions on firms' activities. For their part, business elites after an oil windfall should be opposed to these types of reforms, especially when new entrepreneurs and SMEs pose a credible threat to their business activities. Business elites should try and block the supply of business environment reforms as a means of preserving their rent-derived privileges.

My theory also contains a "demand side" component, which argues that an oil windfall and consequent growth of an oil sector will weaken the *capacity* of nonelites in nonoil sectors to formally engage with the government on policy matters. This means that nonelites are less likely to influence policy making and establish a constructive dialogue with the state on the topic of business environment reforms. In contexts where business associations that address the interests of nonoil sectors or SMEs are present and active, they should have less influence on policy makers' decisions. Policy makers are also likely to punish nonelites who engage in collective action as a way of voicing their demands. This combination of supply- and demand-side forces following an oil windfall means that oil producers should have more restrictive business environments than nonproducers.

3.1 THE IMPORTANCE OF INTERESTS AND AGENCY

The role of *interests* is key to understanding how and why certain policies emerge, and the extent to which the government seeks to regulate the activities of the private sector. A voluminous body of work documents and analyzes how firms, sectors, organizations, or business elites "capture" the policy-making environment and establish an equilibrium that provides them with special benefits and privileges. Reforms that might take away these benefits and privileges are fought tooth and nail. This scenario is a common one that emerges throughout the world—irrespective of a country's level of income or type of government—and occurs with respect to regulatory policies as well as other policies that affect the private sector.

Samuel Huntington's study of the US Interstate Commerce Commission is an oft-cited example of what is termed "regulatory capture."[1] In his study, railroad owners in the United States were able to manipulate a government agency to regulate in a way that supported the interests of the railroads, while ignoring the interests of farmers, consumers, and other logistics providers such as trucking and maritime shipping. For decades, the preferential treatment of railroads remained the status quo until a constellation of forces eventually shifted the equilibrium toward deregulation and the agency was dismantled. To be sure, this is just one example. The evidence of industry-specific interest groups bending regulatory policy to their advantage (and the disadvantage of their competitors) by marshaling finances or promising political support is vast.[2]

Centering on the role of interests—whether at the individual or group level—helps explain the contours of regulation and other types of policies that affect the private sector. Interests are an essential part of the reason why policy reforms occur, and perhaps more often why they *do not* occur. Organized interest groups played a critical role in determining whether or not Latin American countries adopted neoliberal policy reforms.[3] The role of interest groups helps us understand why policy stasis occurs among African cocoa producers.[4] Merilee Grindle explains how interest groups affect the prospects of reform: "[they] use their . . . influence to lock in place or deepen the policies that benefit them, and these interests resist efforts to supersede or revise these policies, even when changed or revised policies would lead to more socially beneficial outcomes."[5]

For both policy makers and private-sector elites, the rent-seeking dynamic shapes their interests about policy reforms. Private-sector elites engage in rent seeking in order to obtain privileges for them or their industry, such as lower regulation or taxation, highly sought-after licenses, or legal restrictions on competition.[6] This is commonly done through lobbying or the offer of financial or political benefits to policy makers. In certain contexts, policy makers are more proactive and demand extortion from industry by using the threat of regulation or adverse policies. Alternatively, they may adopt policies that are specifically intended to punish political opponents and reward supporters. Policy makers and bureaucrats can also benefit from regulation even when there is no specific interest group for them to support. They can generate their own rents when private-sector actors offer them bribes or side payments to overcome steep regulatory barriers.[7]

The role of interests is an important part of the story of policy reform, but agency and the *capacity* of agents can also help explain why policy reform is achieved or not. In places with entrenched rent seeking—such as countries that receive an oil windfall—the prospects of policy reform are shaped by the extent to which segments of the private sector can act as "agents of reform." However, it is often the case that those who are the most supportive of reform are those with the least resources and capacity to influence policy makers. With respect to business environment reforms, the primary beneficiaries of reforms such as new entrepreneurs or small-sized firms tend to have little capacity, at the individual level, to influence the decisions of policy makers. Many assume that large firms have much greater capacity to influence policy makers,[8] a notion that is confirmed in research on the developing world.[9]

This is often attributed to the fact that large firms can marshal more resources toward influencing policy makers, and that the stakes are higher for them having a say in defining policies. Smaller firms have fewer resources and may have less motivation to influence policy makers if the primary benefits of policies will accrue to large firms anyway.[10] Yet, the assumption of a direct, positive relationship between firm size and the capacity to influence policy makers does not always hold. Industry-level conditions, such as the level of competition, and a country's legal and institutional characteristics can render small firms just as influential as large firms. In addition, countries with more developed and diverse political institutions will provide more "entry points" for firms to influence the decisions of policy makers.[11]

Given the relative weakness of any one firm in the private sector (particularly small ones), the role of business associations and sectoral organizations are used to lobby the government on their behalf. In the developing world, there is evidence that better outcomes result when business associations play a part in the policy formation process. A more active business community can help strengthen property rights, reduce information costs, facilitate vertical and horizontal coordination, and generate a number of other benefits for the private sector.[12] Stronger government-business relations at the state-level in India—where the members of business associations are mostly small firms—are linked to higher economic growth rates.[13] However, the challenge of collective action looms large among business associations and there can be downsides to their activities.[14]

Next, I map the interests and agency capacity of policy makers, business elites, and nonelites following an oil windfall. I argue that an oil windfall sets in motion causal mechanisms that directly affect the interests of policy makers and business elites about business environment reforms. I also argue that an oil windfall affects the capacity of nonelites in the private sector to act as agents and demand reforms. An understanding of how these mechanisms operate is important for explaining the trends

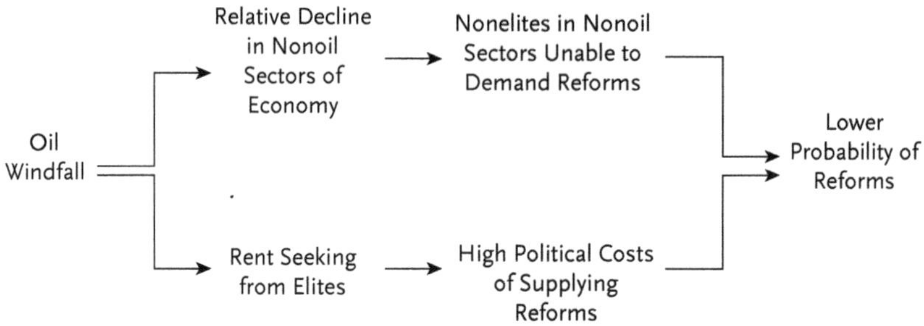

Figure 3.1 Causal Diagram of Link between Oil Windfalls and the Probability of Business Environment Reforms

observed in Chapter 2, as well as in my case studies that appear in the forthcoming chapters. A simplified version of the theory is provided for reference in Figure 3.1.

3.2 INTEREST FORMATION AMONG POLICY MAKERS AND BUSINESS ELITES AFTER AN OIL WINDFALL

An oil windfall transforms government-business relations. The introduction of oil revenues into the government's coffers leads to rent-seeking behavior from business elites and individuals who have close ties to the government. Observing a state flush with a new and abundant source of revenues, business elites petition the government for rents and various forms of preferential treatment. At least initially, leaders of these states routinely appease elites and provide them with rents, sometimes appearing to even encourage the behavior. Quite expectedly, business elites grow dependent on the state's largesse over time.

The types of rents that business elites are seeking tend to be in the form of monopoly rights, no-bid contracts, low-interest loans, or the adoption of preferential regulation and other policies. Rents are intended to raise their profits and limit their competition with others who would want to enter the market and succeed, such as new entrepreneurs. In an important way, the logic of rent seeking hinges on a desire by incumbent, *elite* actors to eliminate the emerging, *nonelite* actors as competitors. Although most work on oil producers glosses over the elite versus nonelite aspect of rent seeking to focus on the precise nature of government-elite relations, my theory breaks down the private sector into these two groups. This approach has shown to be highly useful for building theories about policy making, institutions, and economic outcomes.[15]

Elites want to protect their rents and eliminate their competitors, and this leads them to possess interests that oppose reforms that would lower barriers to market entry and improve access to credit—precisely the opposite of the interests of emerging, nonelite entrepreneurs and firms.[16] Higher barriers to entry, longer bureaucratic delays, and weaker credit institutions are little concern to elites given that they can use political connections to get around them. Unsurprisingly, there is considerable evidence of avoidance and manipulation tactics from elite-owned firms in countries around the world, and there is a distinct disparity in how "favored" versus "disfavored" firms actually experience government regulation.[17]

Policy makers provide rents to business elites after receiving an oil windfall because rents are a way to "buy off" business elites in order to exercise greater decision-making power and/or to acquire their political support. Jill Crystal illustrates this phenomenon by revealing similarities in how leaders in Kuwait and Qatar approached business elites after the onset of oil.[18] Leaders granted rents to influential merchant elites after the oil windfall and, as a result, became less subject to the demands of merchant elites and could exercise greater autonomous decision making. In her words, "Since money was no object, it was easier for the state to buy them out than to repress them [the merchants]. As a result, the oil pact—this tacit trade of wealth for power—emerged."[19]

This tidy exchange between the government and business elites after an oil windfall does not always hold up in exactly the same way over time. Some leaders eventually grow weary of constantly doling out rents, or fiscal crises (such as the oil bust) can encourage them to try and reduce rents to business elites. Kiren Aziz Chaudhry's analysis of Saudi Arabia examines the government's attempts to reduce subsidies to business elites after the oil bust. However, the oil boom era had already led to the rise of a new group of rent-seeking elites who were highly dependent on these rents. Consequently, they were able to hamstring the reform-minded government at the time.[20] In Venezuela, rent seeking reached such an extreme level that a newly elected government in 1989 attempted a "Gran Viraje" (turnaround) and decided to eliminate the rent bonanza. The outcome of this for government-business relations "was a disaster, akin to going from incest to an abrupt divorce."[21]

Given the rise of rent-seeking behavior after an oil windfall and the assumption that business elites are opposed to business environment reforms that strip away or devalue their rents, I argue that this should affect the interests of policy makers about reforms. That being said, we cannot assume that policy makers will always try to placate the interests of business elites and automatically craft a more restrictive business environment. Carefully spelling out the interests of policy makers is critical in order to make predictions about how they will address regulation and other policies that define the business environment.

I start with an assumption that policy makers, prior to deciding to adopt or reject business environment reforms, will attempt to estimate its *costs* and *benefits* on economic and political terms. There is considerable evidence that governments around the world employ some form of cost-benefit analysis to inform policy making, particularly with regard to regulatory issues.[22] However, it might be argued that developing countries, and particularly those with low state capacity, are unlikely to engage in careful cost-benefit analysis to inform policy making. As will be made clear in the following text, policy decisions about the business environment necessitate a relatively low informational requirement—certainly less than what is required when addressing environmental policies, for example.[23] Furthermore, there is evidence that even governments in the developing world can engage in cost-benefit analysis, and likewise in policy areas that may have high informational requirements.[24] Finally, the issue of low state capacity is likely less of a problem for middle-income developing countries as well as those with oil wealth. A recent study finds that natural resource abundance boosts state capacity rather than weakens it.[25]

First, consider the economic costs and benefits of deciding to pursue reforms that would open the business environment and ease restrictions on entry, bureaucratic delays, and access to credit. We know that policy makers in countries that receive an oil windfall have few economic costs and substantial long-term economic benefits to adopting these reforms. Expanding the size of the private sector and reducing dependence on oil will help the country to avoid economic crisis when the price of oil goes down. The sheer presence of abundant oil revenues provides the government with an opportunity to make expensive investments in institutional reforms, or offer training and credit to boost the growth of a new crop of domestic entrepreneurs. Indeed, they have many reasons to do this in light of the stagnation of their nonoil sectors.

Although the fiscal benefits of these type of reforms are negative in the short term, the steady addition of taxable firms into the economy will expand the government's purse over time. The short-term economic costs of reforms should be less of a burden on the budgets of oil producers as opposed to other developing countries that do not have access to and control over a large source of finance. Foregoing reforms will save the government from these expenses in the short term, to be sure, but it sacrifices access to a much larger pool of revenues down the road. In sum, there are *short-term costs* but considerable *long-term economic benefits* to pursuing business environment reforms after an oil windfall.

The political costs and benefits of the decision to pursue reforms are less straightforward. Political costs and benefits are always harder to evaluate and quantify than economic costs and benefits.[26] Consequently, I conceptualize them in terms of a *low* or *high* degree of certainty regarding their value. Also, reforms will affect elites and nonelites in the private sector differently, which will have different political implications for policy makers. And there is good reason to believe that the political costs and benefits of reforms after an oil windfall are measured through their impact on the ability of policy makers to continue "rent seizing," whereby they retain the power to allocate rents to others.[27] After an oil windfall, policy makers reap material and professional benefits from bestowing rents to business elites. As a result, they should be more risk averse when it comes to estimating the political costs of business environment reforms. Rent seizers should always want to remain rent seizers, both now and into the future.

With regard to the impact of reforms on business elites, policy makers should have *high* degrees of certainty about the political costs and benefits of rejecting business environment reforms. Implementing reforms should weaken a policy maker's relationship with business elites (given the latter's interests), and hence they should be reluctant to do this because elites serve as an important source of political strength for policy makers. Elites can mobilize support for the ruling party and use indirect channels of influence to encourage opposition leaders to support them. They can provide direct contributions to policy makers or help finance campaigns in democratic settings. Supporting business elites through rents—and rejecting reforms—is in the self-interest of policy makers as they are then more likely to gain employment in or serve on the board of directors of elite-owned firms after leaving office.

With regard to the impact of reforms on nonelites, policy makers have a *low* degree of certainty about the political benefits of pursuing business environment reforms. To be sure, nonelites in the private sector—such as new entrepreneurs and firms of modest capital—are a substantially larger group than elites and garnering their political support can help policy makers remain in power and continue rent seizing. Business environment reforms can be interpreted as a quasipublic good intended to benefit the welfare of a large segment of the population. But the fact that nonelites are a diverse group—in terms of class, sector, size, and other dimensions—increases the uncertainty about relying on them as a base of political support. Nonelites cannot provide guaranteed political or material benefits to policy makers, and policy makers need to hope that they are rewarded by them at the ballot box or through their quiescence. This is a consideration that especially concerns policy makers who operate in places with an oil windfall, given the unique and substantial benefits that accrue from having the power to allocate rents.

There is also a *low* degree of certainty about the political costs associated with nonelites if policy makers avoid reforms. Higher regulatory barriers inhibit the ability of budding entrepreneurs to enter the market. Restricting access to credit and imposing difficulties to acquire licenses will agitate the numerically large group of nonelites. Nonelites can mobilize against policy makers, use business associations or similar organizations to pressure reforms, or seek to influence the citizenry in order to oust policy makers from office. Nevertheless, nonelites (by definition) are economically weaker at the individual level than elites. Their high degree of diversity should lead policy makers to perceive that they have considerable challenges in lobbying or engaging in collective action.

Overall, a cost-benefit analysis leads us to expect that policy makers in places that experience an oil windfall will adopt interests that are opposed to business environment reforms. The political benefits of supporting business elites (by rejecting reforms) should be perceived by policy makers as higher than the uncertain political costs of not supporting nonelites and the long-term economic costs of rejecting reforms. The political costs of *not* supporting elites (by adopting reforms) should be perceived as higher than the uncertain political benefits of supporting nonelites and the long-term economic benefits of reforms. Further, the perceived value of the economic benefits of reforms should decrease when policy makers have concerns about the stability of the government. Given the advantages that policy makers receive when they continue to rent seize—an opportunity facilitated by an oil windfall—they are likely to discount the long-term economic benefits of reforms and overvalue their short-term costs.

The forces that shape the interests of policy makers and business elites about business environment reforms should remain in place even after a drop in the price of oil. As was observed in Saudi Arabia, the onset of the oil bust did not mean that policy makers gave up rent seizing and elites gave up on rent seeking and just went away peacefully.[28] The pie may be smaller but the actors still want their slice. In a way, a fiscal crisis can strengthen the bonds between policy makers and business elites while raising the intrinsic value of rents, as an economic slowdown makes business elites more concerned about holding on to their slice of the pie. The complete depletion of oil, however, should have its logical effect by raising the prospects of business environment reforms as it would in a place that does not receive an oil windfall. My analysis of the Indian state of Bihar in Chapter 5 tests this implication of the argument. In 2000, Bihar lost its vast coal and mineral reserves because of a redrawing of India's federal boundaries. While a catastrophic event for the state's industrial sector, the elimination of the resource windfall gave way to policies that benefited the growth of local entrepreneurs and SMEs.

3.3 NONELITES AS AGENTS OF POLICY REFORM AFTER AN OIL WINDFALL

The second causal mechanism holds that the declining role of nonoil sectors on relative terms after an oil windfall affects the policy-making influence of nonelite members of the private sector. The government's increasing attention toward expanding the oil sector should lead nonelites to have less collective capacity to formally engage with policy makers. While nonelites already have little political capital at the individual level, we would expect the same at the collective level and with respect to formal channels of influence. Nonelites may retain the capacity to engage in collective action to voice their demands, but this should lower their chances of getting a seat at the policy formation table.

Nascent entrepreneurs and firms of modest capital are the most likely actors to demand that policy makers implement business environment reforms. However, there is good reason to be skeptical about their capacity to effectively demand these reforms. Even nonelites in emerging and innovative nonoil sectors are more likely to own newer firms with less capital and have no political connections. Given the fact that nonoil sectors encompass *everything but oil*, the diversity of firms along class, sector, and size may inhibit coordination around the goal of policy reform. The fact that the prime beneficiaries of policy reform are those with a low capacity to demand them is a familiar scenario. Writing on Latin America, Grindle explains that "those who might benefit from change were generally unorganized, dispersed, politically quiescent, and no doubt unaware of how they might benefit from change."[29]

Even if nonelites are weak on an individual level, they can invoke the powers of business associations or similar organizations to pressure the government on their behalf. While the relative power of business associations may be contingent on broader state-society forces and factors that existed prior to an oil windfall,[30] policy makers may turn toward business associations as partners in policy formation when the hard times set in, such as during an oil bust.[31] That said, in the oil-producing countries that we have knowledge about, most of these associations are guided by elites who have close ties with leaders. In Gulf and Levant countries, it is only in recent years that nonelites have entered elite-dominated sectors of the economy, and it is as yet unknown how business associations will react to their presence.[32] Rent seeking from elites after an oil windfall should bias business associations toward a focus on elite interests at the expense of nonelites. It is telling that an analysis of Kuwait and Jordan found that business associations in oil-rich Kuwait were biased toward elites but business associations in oil-poor Jordan exhibited increasingly inclusive membership starting in the 1970s.[33] Thus, although business associations have played central roles in the reform process in a number of developing countries,[34] they should be weak advocates for the interests of nonelites in countries that receive an oil windfall.[35]

The alternative to business associations as a conduit for influencing policy makers is to engage in direct forms of collective action. Nonelites can use a variety of tactics to pressure the state to adopt reforms, such as by staging protests or coordinating strikes in their sector. The literature suggests that this is a common outcome during the context of economic development, particularly in regard to actors in the informal economy and those whose activities are threatened by mass production, imports, or globalization-like processes. A class-based theory would predict conflict to emerge between nonelites and a government-elite coalition that is built on bonds of rent seeking following an oil windfall.[36] Theories that invoke the concept of dualistic economies with "traditional" and "modern" spheres would hold that conflict between the government and nonelites during the context of economic development is practically inevitable.[37] As oil fuels development, nonelites might rise up to oppose the government as a way of defending a "moral economy."[38]

Although we cannot assume that conflict will automatically emerge between the government and nonelites after an oil windfall, rent-seizing policy makers are unlikely to prioritize the plight of activist nonelites due to the political costs that accompany antagonizing business elites. Instead, policy makers are more likely to punish activist nonelites as a means of preserving a status quo whereby policy makers reap benefits from allocating rents to business elites.

Combining my theory's supply-side component with its demand-side component leads us to the following prediction: countries that receive an oil windfall are more likely to have a restrictive business environment for domestic entrepreneurs and nonelites than countries that do not receive an oil windfall.

3.4 TESTING THE THEORY AND LOOKING AHEAD

I adopt a case-study research design in order to pinpoint the microfoundations of the casual mechanisms that comprise the theory. Testing the theory requires evaluating the evidence of government policy making toward elite and nonelite portions of the private sector; the evidence of rent seeking and the nature of government-elite relations; and the evidence of nonelites' capacity to demand policy reforms. In each of the case studies, I collect quantitative and qualitative data on various aspects of policy making toward the private sector, which includes business regulation, taxation, and credit lending to domestic entrepreneurs and SMEs. Examining data on regulatory, institutional, and tax-related barriers offers insights into the relative ease of opening a business, obtaining licenses and permits, paying taxes, or dealing with other necessary requirements of doing business in a particular context. Data on credit lending and the development of lending institutions provide an understanding of the extent to which the government and financial sector contribute to the growth of domestic entrepreneurs and SMEs. Due to the fact that (1) data availability can vary from country to country, (2) entrepreneurs can face unique barriers in each country and time period, and (3) each country has its own way of categorizing "SMEs," I utilize all relevant data and strive to achieve a holistic understanding of the business environment in each case under consideration.

There are three case-study chapters. Chapter 4 tests the theory in one of the world's major oil producers, Iran. The case study of Iran seeks to reveal how the interests of policy makers and business elites evolved after the oil windfall, and how the capacity of nonelites to demand reforms was affected. I start by examining Iran prior to the oil windfall, when we observe a less restrictive business environment for domestic entrepreneurs and SMEs. I then focus on the period after the oil windfall (1954 to 2014), during which we observe a more restrictive business environment for domestic entrepreneurs and SMEs. Chapter 5 then lays out a comparison of two federal states in India to demonstrate that the theory achieves the following: (1) it can apply to coal-producing states as well as oil-producing countries, (2) it is not sensitive to the type of regime in place (democracy or autocracy), and (3) it can apply to the subnational level of policy making. Chapter 6 then teases out the mechanisms whereby the adoption of business environment reforms is possible in oil producers. I examine the case of Saudi Arabia, one of the weak outliers detected in Chapter 2, and explore how policy makers can reduce the uncertainty surrounding the impact of reforms on business elites' rents, thereby facilitating the introduction of business environment reforms.

Specific types of evidence need to be observed in order to confirm the theory. The causal mechanisms involve interactions between policy makers, business elites, and nonelites in the private sector. One central aspect of interactions between policy makers and business elites is the rent-seeking dynamic that emerges following an oil windfall. While rent seeking is a feature of all economies, oil abundant or not, there needs to be convincing evidence that this behavior *increases* following an oil windfall. There must be evidence of elites petitioning the government for rents and evidence of contracts, licenses, monopolies, credit, or policy adoption that is indicative of preferential treatment for a small group of elites. We would similarly need to see evidence of business elites seeking to protect their rents.

I argued in the preceding text that policy makers protect business elites through granting rents and rejecting business environment reforms. In return, policy makers receive political support that is often accompanied by professional and material benefits. Evidence of business elites serving as a source of political support can be observed, for instance, through reports of loyalty or tacit support of the government's decisions, especially those that are deemed unpopular among the rest of the population. Business elites may provide financial support to specific political parties, help fund election campaigns, or offer public endorsements of policy makers. Examples of direct professional or material benefits can be lucrative positions for retired policy makers (or their family members) at elite-controlled corporations, or bribes and side payments given to public officials in return for preferential policies.

The theory made specific predictions about nonelites in the private sector. Throughout the case studies, nonelites are defined as entrepreneurs and firms of modest capital, or what generally correspond to an SME in the country or state under investigation. Members of the informal economy are included in this group. In light of the cross-national statistical analysis in Chapter 2, there should be case-specific evidence of restrictions on the following: the ability to start a business, access to permits or licenses, access to credit, the ability to file and pay taxes, and regarding other common activities that concern domestic entrepreneurs and SMEs. There needs to be confirmation that these restrictions increased following an oil windfall, or that the loss of a windfall (as is discussed in Chapter 5) results in the decrease of restrictions on entrepreneurs and SMEs.

One way to examine evidence of government regulation is to look at individual- or firm-level data. However, this approach can be problematic given the subjectivity that can surround some aspects of regulation. Chapter 2 dealt with this issue to a certain extent by analyzing cross-national quantitative measures of regulation, but it would be fairly easy to craft a case study that relies on anecdotes from entrepreneurs and firm owners who recite a litany of complaints about regulation and governmental interference in their activities. We can find such individuals in any country, oil abundant or not. For this reason, I triangulate sources of evidence by relying on quantitative and qualitative data.

The theory argued that an oil windfall blunts the capacity of nonelites to effectively demand business environment reforms. Business associations, following an oil windfall, should become relatively weak representatives for the interests of nonelites. Although we know that historical and societal factors affect the ability of business associations to influence policy, it is important to examine their activities in contexts where they are relevant. Here again, I seek to combine multiple sources of evidence. Relations between policy makers and nonelites are also argued to be generally poor following an oil windfall. Nonelites may attempt collective action when policies are perceived as exclusively benefiting elites, and evidence of mobilization efforts can be strikes or petitions to the government. We expect policy makers in the context of an oil windfall to punish activist nonelites, and prevent them from playing a role in policy formation either through punitive actions, higher regulation, or marginalization of business associations that represent their interests. For certain, nonelites are never completely stripped of all power to challenge a government that receives an oil windfall—even in dramatic ways, as discussion of the 1979 Iranian revolution in Chapter 4 reveals—but they should be unable to alter the general contours of policy making toward the private sector. I now turn to the case of Iran.

CHAPTER 4
Oil and the Business Environment in Iran

This chapter provides an in-depth case study of Iran, a major producer of oil and a founding member of OPEC. Oil was first discovered in Iran in 1908, but the resources were initially controlled by the Anglo-Iranian Oil Company (the company that later became British Petroleum). The Iranian government did not receive a significant stream of revenue from its oil until 1954, following an oil nationalization crisis. From 1955 to 1962, Iran's oil revenues were more than three times the value of the country's total foreign exchange receipts (oil plus loans) during the previous two decades.[1] By 1973, Iran's oil revenues reached five billion US dollars. By 1977, they were twenty billion US dollars (see Figure 4.1).[2]

Given the theory proposed in Chapter 3, government policy making in post-1954 Iran (after the windfall) should be characterized by the imposition of many government regulations and restrictions over the activities of domestic entrepreneurs and SMEs. By contrast, policy making in pre-1954 Iran (before the windfall) should be characterized by comparatively fewer regulations and restrictions. The case-study logic is thus a "before-after" research design, whereby a longitudinal case is divided into two subcases.[3] In order to construct a case study of Iran, I rely on a wide range of primary and secondary source materials from academic research to newspapers to time-series data on manufacturing licenses.

Iran is an appropriate case for testing the book's theory. The country lies squarely "on the regression line" and thus is not an atypical oil producer with an unusually restrictive business environment given its level of oil income and other variables. This is shown by taking the regression

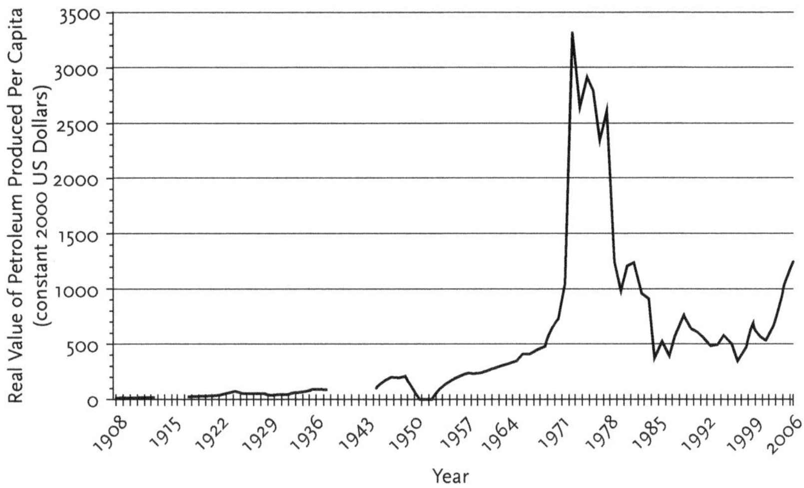

Figure 4.1 Iran's Oil Income, 1908–2006

equations from Chapter 2 and generating predicted versus actual values for Iran using the full set of control variables. In the area of start-up regulation, Iran's predicted value is 5.29 and its actual value is nearly identical at 5.27. The predicted and actual values are close, but somewhat less so, for permits regulation (8.17 and 9.23, respectively). However, the predicted and actual values for tax regulation (8.32 and 8.93) and contracts regulation (10.35 and 9.91) are both very close. Other variables produce similar results.

Iran is also an inherently interesting case given the fact that its oil windfall spans across two political eras separated by the 1979 revolution. Unlike many oil producers, such as the Gulf Arab states, Iran experienced a political transition from monarchical rule after it was already an oil producer. Consequently, if we find that the oil windfall has affected Iran's business environment across two different political systems we should be more confident that the theory holds on a broader level.

I break up my analysis into chronological periods and within each period examine the evidence for the microfoundations of the theory. I begin with Iran's preoil era during the Qajar and Pahlavi dynasties. Acquiring an understanding of the business environment in the preoil era is an important part of testing the theory. Indeed, even if Iran's business environment after the oil windfall was highly restrictive, we have no way of knowing if this was any different from what existed prior. Without an understanding of the preoil era, we cannot know if the oil windfall changed anything at all about government policy making. After providing an overview of the preoil era, I focus on policy making under the government of Shah Mohammad Reza Pahlavi between the years 1954 and 1979, when oil revenues first became a substantial source of income for the nation. Following this, I center on policy making after the 1979 revolution and until the early 2010s.

At many times in the chapter, I refer to Iran's *bazaar*,[4] which is a group of economic actors that is important to define before moving ahead. The bazaar comprises hundreds of thousands of entrepreneurs, traders, artisans, shopkeepers, and owners of small factories or workshops who are mostly (but not entirely) of modest capital. The bazaar's activities encompass a sizeable share of the Iranian economy. By 1979, the bazaar controlled up to two-thirds of retail trade, one-third of imports, and 97 percent of total domestic manufacturing, and also absorbed 68 percent of manufacturing labor.[5] The capital city of Tehran is home to the country's largest and main bazaar, which, by the mid-1970s, included twenty thousand to thirty thousand commercial units and controlled roughly 20 percent of total market volume.[6] Beyond the capital, regional bazaars are the beating heart of local economies.[7]

My case study demonstrates that Iran's oil windfall led to accelerated rent seeking from business elites and a reduction in the capacity of domestic entrepreneurs and SMEs (in particular those located in the bazaar) to formally engage with policy makers on the topic of reforms. This resulted in a business environment that imposed various barriers to the activities of domestic entrepreneurs and especially small-sized firms. Consistent with my theory, the postrevolutionary government's approach to the business environment did not markedly change with respect to my key variables of interest. Forces put in place after an oil windfall remained in place even beyond the threshold of a revolution.

4.1 BEFORE OIL: POLICY MAKING AND THE PRIVATE SECTOR UNDER THE QAJARS AND PAHLAVIS

In the early years of the Qajar dynasty (r. 1785–1925), merchants had a "good working relationship" with the government and they exercised autonomy on a number of levels.[8] They basically traded cash for their economic freedom, and in return the government established a business environment that was generally supportive of their activities. Merchant- and artisan-based guilds at the time were autonomous, self-regulating, and well-organized, and also played a central role in the development of Iranian society.[9] Merchants held active and important roles in politics and policy making, utilizing the power of guild organizations and capitalizing on their social ties to clerics. During this period, the strength of merchants on economic and political terms was "greater in Qajar Iran than elsewhere in the Middle East."[10]

Since the days of the Silk Road, Iran's geographic proximity between Europe and Central Asia has helped to foster the growth of overland trade routes. Thus, a well-established network of commercial trade already existed by the time the Qajars came to power in the late eighteenth century. In urban centers, merchants were engaged in trade and handicraft production while artisans and small workshops outside the cities produced for local markets.[11] Merchants who operated outside of urban centers exercised considerable autonomy from the government. The Qajar administration was highly decentralized both by necessity and due to its recognition of the importance of regional interests. This benefited nonurban merchants because local authorities retained fiscal autonomy from the central government, controlled surplus revenue, and dictated economic decision making in their region.[12] The government appointed chief merchants (*malek al-tojjar*), who wielded more power than provincial bureaucrats and often served as advisers to the shah.[13]

This state of affairs lasted until the 1830s. Throughout the next several decades, merchants would come to face increasingly difficult business conditions. Yet, it is important to emphasize that the cause of this change was *external* to the Iranian government. After a series of military losses, the Qajars were forced to grant special economic rights to foreign nations. In 1828, Russia prevailed over Iran in a territorial dispute, and consequently the Qajars were forced to lower tariffs on Russian goods and grant exemptions to imports. In 1856, Iran's second invasion of Afghanistan was defeated by the British and the Qajars were forced to grant the British commercial privileges. Western companies now had a distinct advantage over Iranian merchants, and the rate at which British and Russian goods flowed into the country was rapid. Both foreign nations saw their presence in Iran as "absolutely vital" to their broader imperial goals,[14] and some scholars describe Iran as a "semicolony" during this period.[15]

In response, Iranian merchants pressured the government to intervene on their behalf.[16] Initially, most of their requests went unfulfilled. But this was not because the government was indifferent to the needs of merchants. In fact, "there was little the government could do, short of risking war with the Western powers."[17] Nevertheless, Mohammad Shah (r. 1834–1848) tried to stop the use of European cloth among his courtiers, and some of the merchant-led boycotts on Western goods were ultimately successful.[18] This sparked an era of increasing mobilization and politicization among Iranian merchants, and saw coordination among handicrafts producers and small manufacturers across the country.[19] As more concessions were granted during the mid-to-late 1800s, the conditions for merchants further deteriorated, but they became united around the goal of opposing foreign influence. This is observed in the response to a tobacco concession that was granted to a British company in 1890. Merchants and clerics lobbied the government and organized a boycott, and the government canceled the concession two years later.[20]

As a whole, the years between 1830 until the end of World War I were marked by difficult business conditions for most Iranian merchants. However, foreign influence produced these conditions and the Qajars were a state with its hands tied. Although the Qajars certainly benefited in financial and geopolitical terms from helping to accelerate Western influence in Iran, they positively responded to merchant demands when they were pressured. The ability of merchants to reverse important concessions on more than one occasion suggests that they had the capacity to penetrate the Iranian state and force changes in policy.

The Qajar dynasty ended because of a military coup, led by Reza Khan, which forced the removal of Shah Ahmad in 1925 and initiated the beginning of the Pahlavi dynasty. The nation and economy that Reza Khan (now known as "Reza Shah") inherited was on the verge of collapse. Fortunately for the state, oil revenues slightly increased in 1931 and remained at this level throughout the duration of Reza Shah's rule. The increase in oil revenues—along with a desire to mimic the policies of Atatürk in neighboring Turkey—led Reza Shah to embark on a reform program that established the foundations for industrial development and spurred the growth of a truly *domestic* private sector in Iran. Reza Shah eliminated many of the commercial treaties and concessions adopted under the Qajars, seeking to break the legacy of foreign influence that defined Iran's economy for so long.[21]

Reza Shah's government established an important role for the country's domestic private sector—including small-sized capital—and laid the foundation that would be essential for improving the business environment for Iranian entrepreneurs. He fostered the growth of small-scale manufacturing in Iran, which was primarily fueled by private-sector actors and not the state. During the 1920s, the more successful firms tended to be small and domestically-owned, whereas large-scale manufacturing owned by the state or foreign investors largely failed. The new firms—such as soap, glass, and textile manufacturers—were primarily small workshops geared toward local demand and utilized unskilled and artisan labor. [22] Importantly, many of the entrepreneurs who emerged at this time came from "humble origins" in the bazaar.[23] That being said, bazaaris who were more upwardly mobile, or defined as middle class with respect to income, were better positioned to take advantage of these opportunities than those at lower rungs of the income ladder.

Reza Shah's government made advances in the area of institutional development, which was necessary for establishing a more competitive business environment in Iran. Ministries of the economy were founded and divided into sectors to regulate agricultural, commercial, and industrial activities. Private property was institutionalized and the government required the registration of stock companies and established regulations in 1931. A commercial code was passed in 1932.[24] In 1936, the formal procedures for registering a business were codified. A number of specific policies were intended to encourage growth in the domestic private sector. For instance, entrepreneurs could import capital goods and half-finished goods duty free.[25] Other exemptions, rebates, and credit from the Agricultural and Industrial Bank (established in 1933) were also provided. The government systematically discouraged the role of foreigners in the development process.[26]

To be sure, not all of Reza Shah's policies directly benefited the domestic private sector. However, many of them can be explained by considering the historical context. After the start of the global depression in 1929, the government established trade monopolies on certain goods, such as wheat, and later made barter agreements with Germany and the Soviet Union.[27] Internal tariffs on domestic trade were imposed,[28] which no doubt affected many merchants. Some policies initially had a positive effect that arguably hurt the private sector over time. One example was the elimination of taxes on 216 guilds. Although eliminating taxes was a positive outcome from the perspective of guild members, this took away the power that craft and trade masters had over apprentices, artisans, and employees—and was one way of weakening the highly autonomous guild system.[29]

Overall, Reza Shah made significant progress toward the adoption of policies and institutions that helped to expand the role of the domestic private sector in Iran's economy. His government emphasized domestic over international capital; inaugurated many formal institutions that fostered a more competitive business environment; supported the small-scale private sector; and refrained from overt antagonism of merchants, artisans, and other members of the bazaar. For decades prior to his rule, Iranian merchants and artisans were clamoring for a business environment in which their interests were prioritized, and in which local entrepreneurs could grow. In many ways, Reza Shah gave them what they wanted. Under his government, aspiring entrepreneurs and middle-class merchants became important actors in Iran's quest for economic development.[30]

Reza Shah's rule came to an end with World War II. He was forced by Western governments to abdicate the throne and step aside for his son, Mohammad Reza, who at the time was just twenty-one years old. After assuming power, Mohammad Reza Shah pursued the same economic path started by his father, although his government was initially preoccupied with a Western occupation in the north and separatist movements. The state's priorities shifted again in 1949 as a major domestic crisis started to unfold. The Iranian public began pressuring the government to nationalize the oil industry, a movement that was led by Mohammad Mossadegh. In 1951, Mossadegh became prime minister and a rivalry ensued between him and Mohammad Reza Shah. This culminated in a coup orchestrated by the US and British governments, which removed Mossadegh and placed Mohammad Reza Shah firmly in power. The significance of this event is beyond the scope of this book,[31] but one key outcome was a renegotiation of the oil contract and the formation of a consortium. By 1954, the Iranian government started to receive a substantially higher share of profits from its oil production. Iran officially receives its oil windfall.

To summarize Iran's prewindfall era, there was variation in the domestic business environment over the course of Qajar and early Pahlavi rule. Prior to the 1850s, Iranian merchants operated in a largely autonomous and self-reliant manner. The government provided the business sector with a high degree of freedom, consistent with what the theory predicts given that there was no oil windfall at the time. Although the period between 1850 and 1921 signals a more restrictive business environment, we know that foreign influence—not the government's impetus—was the primary cause. In fact, the government made concessions to Iranian producers on important occasions, even in the face of external pressures not to do so. Once Reza Shah took over, there was a distinct emphasis on the role of the domestic private sector in the nation's quest for industrialization. The government quickly adopted policies and institutions that proved essential to the growth of small-sized firms and the Iranian private sector as a whole. However, as the next section demonstrates, after Iran receives its oil windfall any momentum toward creating a business environment that supports small-sized firms and entrepreneurs comes to a stop.

4.2 THE OIL ERA: POLICY MAKING AND RENT-SEEKING ELITES DURING THE WHITE REVOLUTION

After the coup and renegotiation of the oil contract, Iran saw a rapid increase in its oil revenues and a flood of foreign capital, which led to an immediate boom in investment.[32] However, any optimism this may have generated about the strength of Iran's postcoup economy would evaporate in just a few years. By 1959, the country was on the verge of an economic crisis. The major issues at the time were the expensive commitments to industrialization, foreign debt, a bloated military, and rising inflation. In spite of the grand hopes that came with increased oil revenues—alongside the progress made under Reza Shah—it was clear that economic development in Iran would prove more difficult than imagined. The reforms suggested by the International Monetary Fund worsened the situation. The government shut off access to luxury goods, increased tariffs on nonessential goods, restricted bank credit, and banned the sale of foreign exchange.[33] Politically, Mohammad Reza Shah struggled with military officials, civilian leaders, and the elites whom he allied with after returning to power in the coup.[34] So far, Iran's oil era was defined by economic and political uncertainty.[35]

In the midst of this, Mohammad Reza Shah introduced a comprehensive program named "The White Revolution between the Shah and the People." The government laid forth ambitious goals in the areas of industrialization, economic and social modernization, and agrarian reform. Industrial plans were to increase support for consumer-based industries, boost profit-sharing, and sell state-owned factories (the latter to finance land reforms). Socially and educationally, Mohammad Reza Shah brought voter enfranchisement to women and introduced a literacy corps.[36] Agrarian reforms, billed as the cornerstone of the program, would hopefully increase agricultural productivity, confer millions of acres to landless peasants, and achieve other goals. In reality, the land redistribution program was a failure. Half of all land was still controlled by large landlords, half of the landless peasants received nothing, and the other half of landless peasants received plots too small to provide for their families.[37]

In spite of failed agrarian reforms, the White Revolution's industrial reforms coupled with increasing oil revenues resulted in a period of impressive economic growth. From 1963 to 1978, Iran's GDP grew at an annual rate of almost 11 percent—a rate that was nearly unmatched in the world at the time.[38] Per capita income jumped from 166 US dollars in 1953 to 1,514 US dollars in 1977. Steady progress was achieved in the development of industry. The number of industrial firms in Iran multiplied from fewer than 1,000 to about 6,200 from 1957 to 1974.[39] Indeed, the country experienced a "minor industrial revolution" during these years.[40]

Alongside impressive industrial growth came the rise of rent-seeking behavior from Iranian business elites. Rent seeking was spurred on by the oil windfall and supported by the government through its expressed desire to expand Iran's nonoil sectors. After the oil windfall, there were many incentives for individuals with ties to Mohammad Reza Shah's government to seek out rents. The rents that elites received came in the form of monopoly rights; subsidies; free or cheap credit; free or cheap industrial inputs and raw materials; and generous tax breaks.[41] Many who received rents were individuals with close personal ties to the government or members of Mohammad Reza Shah's family. Their close proximity to the throne allowed them to solicit rents from a monarch who seemed more than happy to hand them out.

As per Mohammad Reza Shah's industrialization plans, the country's private sector was tasked with handling light industry while the state would handle heavy industry. Distributing rents to private sector elites was thus justified by the leader as a way of encouraging the private sector to take part in industrialization. In reality, the role of the state in the economy far surpassed that of the private sector. During this period, the government gained or maintained control of oil, mining, modern manufacturing enterprises, most banking, insurance, transport, communications, and many farms and agribusinesses. Unsurprisingly, state monopolies were one of the major grievances the private sector had with the government at the time.[42] Even as oil revenues were steadily increasing, the ratio of private to public investment declined from 1.53:1 in 1959 to 0.76:1 in 1972.[43]

Unlike his father, Mohammad Reza Shah's approach to industrialization exhibited a distinct bias for large scale over small scale. Any promotion of small-scale industry from 1962 to 1972 was "an exception to the rule."[44] Large-scale projects were plied with rents that helped them garner huge profit margins while "little was done for those on the bottom rungs of the economic scale."[45] This was in spite of the fact that, as of 1967, the contribution of small-scale industries to national income was roughly equal to that of medium- and large-scale industries.[46] Rent seeking and a large-scale bias were observed in the agricultural sector as well—despite the White Revolution's stated goal of fostering smaller-sized units. Although large-sized units (more than fifty hectares) represented less than 1 percent of all agricultural units, they "enjoyed the full support of the state" with subsidies, tax breaks, and low interest credit. For small-sized agricultural units, the state "erected many barriers to their viable operation." Additionally, state irrigation policy and credit institutions ignored the needs of small-sized units.[47]

The government's approach to credit lending showcases the power of rent seeking during this period, and highlights the numerous disadvantages that nonelite entrepreneurs and firms faced.[48] For certain, if

Mohammad Reza Shah wanted the country's private sector to play any role in economic growth his government would have to address the issue of credit. Establishing credit institutions and adopting policies to inform lending practices were therefore essential tasks of the White Revolution era. The volume of credit was directly tied to the oil windfall. Massoud Karshenas explains, "In particular in the case of Iran, where the government supplied the bulk of the funds to industrial banks from its oil surpluses, control of finance was an important potential source of controlling the pace and structure of private investment."[49] Access to credit became a primary way that the government influenced the growth of industrial projects and the very nature of development in Iran. But the role of credit also became an arena of rent seeking and consequently strengthened the bonds between the state and elites, while sharpening the divide between private-sector elites and nonelites.

Many of the banking institutions used to distribute credit were established just prior to or during the White Revolution, such as the Industrial Credit Bank (est. 1957), the Industrial and Mining Development Bank of Iran (est. 1959), and the Development and Investment Bank of Iran (est. 1973). These banks made investments and loans that totaled in the billions of dollars, and interest rates were substantially lower than what the private sector was accustomed to receiving.[50] Whereas interest rates typically ran between 12 to 30 percent for short-term loans, the government established rates between 6 and 8 percent for medium- and long-term loans.[51] These rates tended to be lower than what was available from bazaar-based lenders.[52]

Mohammad Reza Shah's government exercised complete control over credit-lending decisions. This meant that those who had preferential access to the monarch could exploit this for their own benefit, which "opened the door to varied forms of rent-seeking behavior and corruption."[53] Indeed, "[c]apital was not scarce in Iran during the late Pahlavi era; however, it was highly politicized. . . . Creditworthiness was based on political and cultural allegiance to the Pahlavi monarchy and specifically to its modernizing agenda."[54] Credit allocation was one of the main ways that Mohammad Reza Shah solidified his connections with devoted elites whom he sought to rely on for political support. Yet, credit "largely bypassed the vast majority of medium-sized and smaller (typically labor-intensive) manufacturers. . . ." Among these manufacturers were those located in the bazaar, most of whom were systematically deprived of credit.[55] Overall, credit-lending practices at the time provided cheap credit to "the small dominant class . . . while denying it to the middle and poor segments of the population."[56]

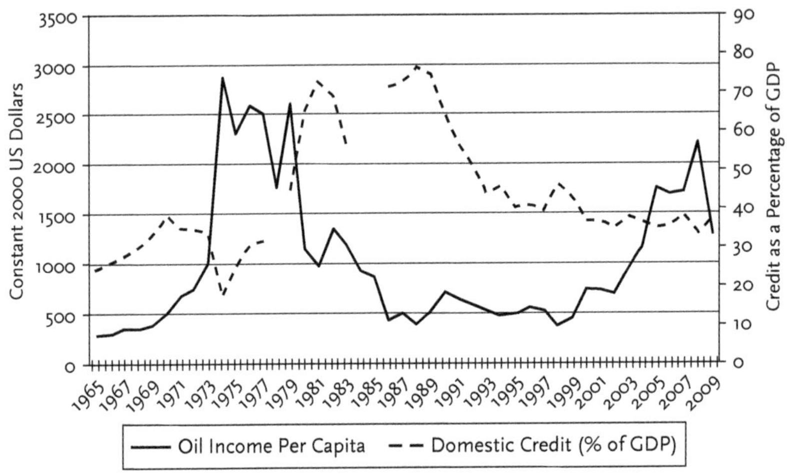

Figure 4.2 Iran's Oil Income and Domestic Credit Provided by Banking Sector, 1965–2009

Figure 4.2 displays a plot of Iran's oil income alongside domestic credit provided by the banking sector as a percentage of GDP from 1965 to 2009.[57] In the initial years of the White Revolution, domestic credit increases in tandem with oil income. However, by 1970, credit from the banking sector begins to decline overall even while oil income is rapidly increasing. Iran's domestic credit as a percentage of GDP reaches its lowest point at the start of the oil boom. From 1965 to 1977, there is an inverse relationship between oil income and domestic credit (the correlation is -0.431). Similar to what was observed in the statistical tests in Chapter 2, more oil is associated with less credit lending.

Another way that Mohammad Reza Shah's government restricted the growth of nonelites in the private sector was in the area of industrial licenses. The government established strict requirements for licenses during the White Revolution, justified as a way of focusing investment on priority sectors, restricting industry to beyond 120 kilometers of Tehran, and preventing "an excessive number of small plants."[58] Although the geographic restriction was explained as a way of discouraging "fly-by-night new enterprises,"[59] this had an obvious effect on entrepreneurs located in the Tehran bazaar. While most of these individuals were denied credit and unlikely to obtain a license anyway, the geographic restriction was an additional disincentive to pursuing one. However, a license provided an entrepreneur with subsidized water and electricity, financing from development banks, tax exemption status for ten years, and a number of other benefits.[60] Many new industrialists who received licenses went on to reap huge profit margins, reportedly as high as 50 to 200 percent.[61] Predictably, this led to an extreme concentration of wealth in the top echelon of Iran's private sector. By the 1970s, it was estimated that just ten families controlled 78 percent of the largest industrial enterprises in Iran.[62] A common expression at the time was that the entire nation's wealth was controlled by a "family of a thousand people" (*khanevadeh hezar nafar*).

The restriction of industrial licenses is further evidenced by analyzing time-series data on establishment and operations permits to new manufacturing firms. As Figure 4.3 displays, the number of permits provided under Mohammad Reza Shah's government was increasing during the White Revolution but still represented a tiny fraction of the permits given out throughout the last sixty years.[63] Although a distinction between public versus private ownership complicates a straightforward understanding of license granting after the revolution (discussed more in the following text), the main observation is that, under Mohammad Reza Shah, license granting was very modest in spite of the fact that these years overlapped with the first oil boom. Overall, there is an inverse relationship between oil income and the number of establishment permits granted from 1960 to 1977 (the correlation is -0.28).

An arguably better way to investigate the relationship between oil income and permits is to use regression analysis. I run a negative binomial model using maximum likelihood estimation and robust standard errors with the number of permits as the dependent variable. The independent variables are oil income, Iran's *Polity* score of regime type, population

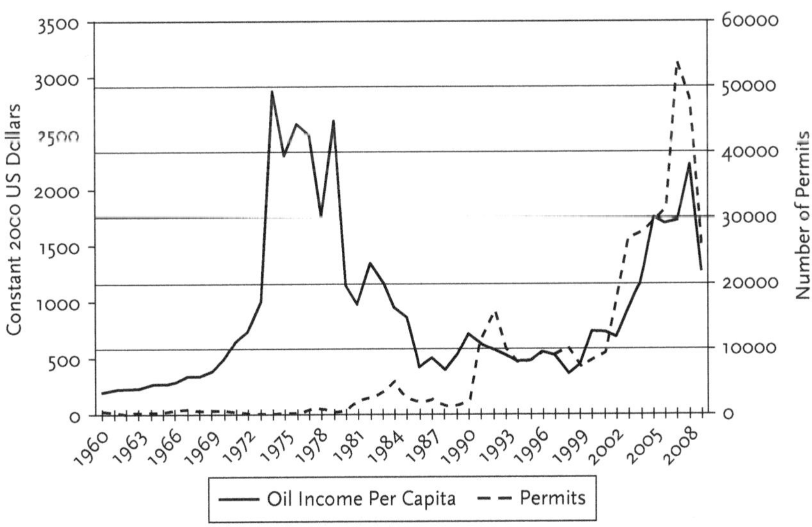

Figure 4.3 Iran's Oil Income and Manufacturing Permits, 1960–2009

Table 4.1 OIL AND OPERATION PERMITS FOR NEW MANUFACTURING FIRMS IN IRAN, 1960 TO 2008

Independent variables	(1)	(2)	(3)	(4)	(5)	(6)
Oil Income	-0.0002**	-0.0002***	-0.0002***	-0.0002***	-0.0002***	-0.0002***
	(0.0001)	(0.0001)	(0.0001)	(0.0001)	(0.0001)	(0.0001)
Population (log)	3.7***	3.8***	4.232***	4.003***	3.799***	3.982***
	(0.216)	(0.319)	(0.354)	(0.37)	(0.32)	(0.444)
Regime Type	—	-0.003	-0.004	-0.001	-0.003	-0.003
	—	(0.017)	(0.016)	(0.017)	(0.017)	(0.017)
Change in GDP pc	—	0.0003***	0.0002	0.0002**	0.0003**	0.0003**
	—	(0.0001)	(0.0001)	(0.0001)	(0.0001)	(0.0001)
White Revolution	—	—	0.413	—	—	—
(1962–1963	—	—	(0.253)	—	—	—
to 1976–1977)						
Post-1979	—	—	—	-0.195	—	—
Revolution	—	—	—	(0.248)	—	—
(1979–1980						
to 2007–2008)						
Iran–Iraq War	—	—	—	—	-0.009	—
(1980–1981	—	—	—	—	(0.123)	—
to 1987–1988)						
Post-Iraq War	—	—	—	—	—	-0.161
(1988–1989	—	—	—	—	—	(0.258)
to 2007–2008)						
Constant	-58.281***	-60.107***	-67.845***	-63.531***	-60.091***	-63.233***
	(3.856)	(5.716)	(6.34)	(6.488)	(5.734)	(7.81)
Observations	47	46	46	46	46	46
Years of Data	1960–2008	1960–2008	1960–2008	1960–2008	1960–2008	1960–2008

The results are from a negative binomial regression using maximum likelihood estimation and robust standard errors. The dependent variable is the number of operation permits issued in a given year for newly established manufacturing firms. All independent variables are lagged one period. Variables indicating time periods or events are dummy coded. * sig at 0.1; ** sig at 0.05; *** sig at 0.01.

(logged), the annual change in GDP per capita growth, and dummy variables to indicate time periods and important events. All independent variables are lagged one period and the years of available data are from 1960 to 2008. The results appear in Table 4.1 and Table 4.2, indicating that oil income is negatively associated with the number of operation permits and establishment permits at $p < 0.01$ levels of statistical significance. Dummy variables for the White Revolution or other time periods are not significant predictors of the number of permits granted. These results show that,

Table 4.2 OIL AND ESTABLISHMENT PERMITS FOR NEW MANUFACTURING FIRMS IN IRAN, 1960 TO 2008

Independent variables	(1)	(2)	(3)	(4)	(5)	(6)
Oil Income	-0.0004***	-0.0003***	-0.0004***	-0.0003***	-0.0004***	-0.0003**
	(0.0001)	(0.0001)	(0.0001)	(0.0001)	(0.0001)	(0.0002)
Population (log)	4.219***	4.329***	4.08***	3.951***	4.369***	4.29***
	(0.219)	(0.368)	(0.541)	(0.479)	(0.384)	(0.718)
Regime Type	—	-0.005	-0.006	-0.013	-0.01	-0.005
	—	(0.031)	(0.032)	(0.032)	(0.033)	(0.032)
Change in GDP pc	—	0.0001	0.0002	0.0001	0.0001	0.0001
	—	(0.0001)	(0.0002)	(0.0001)	(0.0001)	(0.0001)
White Revolution	—	—	-0.265	—	—	—
(1962–1963 to 1976–1977)	—	—	(0.381)	—	—	—
Post-1979 Revolution	—	—	—	0.406	—	—
(1979–1980 to 2007–2008)	—	—	—	(0.363)	—	—
Iran–Iraq War	—	—	—	—	0.133	—
(1980–1981 to 1987–1988)	—	—	—	—	(0.244)	—
Post-Iraq War	—	—	—	—	—	0.032
(1988–1989 to 2007–2008)	—	—	—	—	—	(0.509)
Constant	-65.93***	-67.95***	-63.486***	-61.643***	-68.7***	-67.28***
	(3.819)	(6.575)	(9.686)	(8.312)	(6.858)	(12.429)
Observations	47	46	46	46	46	46
Years of Data	1960–2008	1960–2008	1960–2008	1960–2008	1960–2008	1960–2008

The results are from a negative binomial regression using maximum likelihood estimation and robust standard errors. The dependent variable is the number of establishment permits issued in a given year for newly established manufacturing firms. All independent variables are lagged one period. Variables indicating time periods or events are dummy coded. * sig at 0.1; ** sig at 0.05; *** sig at 0.01.

throughout the period from 1960 to 2008, oil income was negatively associated with the number of industrial permits granted.

The government's strict requirements about who could receive a permit or access credit disadvantaged those with little start-up capital and those who came from the bazaar. For the private-sector elites who had connections to Mohammad Reza Shah's government, however, tremendous gains were made. This resulted in "strong clientelistic ties that reinforced and perpetuated the rent-generating activities of the state,"[64] and led to the rise of a "parasitic element grafted on the monarchy."[65] Nonetheless, private-sector elites generally did not play a role in shaping economic policy during this time. Policy making was highly centralized in Mohammad Reza Shah's government and he "felt no need to consult the private sector."[66] In spite of their close ties to the monarchy and the abundance of rents they received, private-sector elites were "not allowed to participate in politics or to organize autonomous organization."[67] Business associations were weak. The government merged the Chamber of Commerce with a new Chamber of Industries and Mines to create a single chamber that "had absolutely no input into the planning process and the economic strategy of the government." Mohammad Reza Shah was reportedly irritated when a few industrialists won seats in the Senate and Majles (Parliament) during the mid-1970s.[68]

Even though private-sector elites were unable to shape policy making, Mohammad Reza Shah's government continually provided them with rents as a way to bolster his rule. This is best exhibited in his dealings with the Ministry of Economy, as detailed in a study by Vali Nasr. Formed in 1963, the Ministry of Economy was critical to Iran's rapid industrial growth during the White Revolution. However, members of the ministry wanted to increase their autonomy and tried to work against the interests of politically influential, rent-seeking elites. Mohammad Reza Shah acted swiftly to stop this from happening, and by 1969 he broke down the ministry's influence. In essence, he "wanted economic policy-making to accommodate patrimonialism" and expected it to "be more flexible and to accommodate powerful interest groups."[69] By granting rents and responding to powerful interests, he was seeking to strengthen his rule and rely on these elites as a core part of his "new basis of support."[70] Consequently, he had no choice but to terminate an organization that was seeking more autonomy and could have prevented him from converting the oil-funded rents into political support. Nasr explains, "petrodollars gave him the impression that he could choose political interests over institutional change, and that he could preserve the patrimonial basis of his authority without adversely influencing the pace and course of development."[71]

The evidence presented in the preceding text shows that Iran's oil windfall led to rampant rent seeking from private-sector elites, whose needs were satisfied as a means of establishing a base of political support for Mohammad Reza Shah. This dynamic is most visible during the White Revolution, which was Mohammad Reza Shah's first opportunity to introduce a comprehensive policy program after returning to power in a coup. Although the specific goals of the White Revolution were certainly informed by the economic and political problems the government faced at the time, the fact that Iran had received its oil windfall was an important force in shaping how the government actually dealt with these problems. Oil allowed the government to provide generous rents directly to private-sector elites during the White Revolution, which was a way to cement political support on which to better solidify Mohammad Reza Shah's rule. Policy making toward the private sector therefore had two primary goals. The first goal was to benefit handpicked elites who were supporters of the monarchy. The second goal was to intentionally prevent nonelites, such as those in the bazaar, from carving out their own slice of the pie and capitalizing on the numerous economic opportunities at the time. But nonelites also faced many other government-imposed restrictions during these years that inhibited their capacity to effectively demand a more supportive business environment, a subject to which I now turn.

4.3 THE OIL ERA: NONELITES AND DEMAND-SIDE FORCES DURING THE WHITE REVOLUTION

Policy making during the White Revolution had a major impact on Iranian entrepreneurs and firms, especially those located in the bazaar. Preventing bazaaris from accessing credit and licenses were not trivial policy decisions because the bazaar represented the vast majority of Iran's actual pool of entrepreneurs and economic actors. These policies should be understood as the result of the oil windfall. Oil fostered rent seeking from private-sector elites, leading the government to adopt policies that supported this tiny group while restricting others from similar opportunities. Granting rents to elites was a way that Mohammad Reza Shah sought to solidify his base of support and bolster his rule. Policy making after 1954 was different from what existed prior to the oil windfall. While Reza Shah tried to support upwardly mobile bazaaris and boost the growth of small-sized firms, Mohammad Reza Shah adopted a distinctly hostile approach toward them. Under the younger Pahlavi, "the whole bazaar ... [was] becoming politically and economically marginalized."[72] This made it increasingly difficult for the bazaar to demand that the government provide a more supportive business environment. They were shut out of the policy-making arena and, in the end, had to rely on collective action to voice their concerns. Their mobilization efforts coincided with a broader movement that became the 1979 revolution.

The economic growth associated with the White Revolution's reforms affected the bazaar in different ways. On the one hand, policies that introduced modern manufacturing and factory production had a negative effect on bazaar artisans and workshops as it did in other developing countries at the time. For instance, in 1965, the government decided that it wanted to boost Iran's shoe industry by opening several new factories, banning shoe imports, and raising tariffs on leather. As expected, this had a highly negative effect on the bazaar's role in shoe production.[73] On the other hand, trade and retail-oriented sectors of the bazaar benefited due to increasing consumerism and urbanization that accompanied rapid economic growth.[74] In Tehran, some bazaar-based retailers branched out and opened shops in new shopping districts that were modeled after Western-style malls. Although the new shopping districts were actually opened by the government with the purpose of driving business away from the bazaar, wholesalers based in the bazaar often provided goods and storage.

Government-sponsored shopping malls were just one way of cutting into the bazaar's slice of the economy. By the mid-1970s, bazaaris would become "increasingly under siege" as the government enacted an aggressive antiprofiteering campaign.[75] From 1972 to 1976, inflation increased and retail prices rose from an index of 75 to 120. The prime causes of inflation were numerous,[76] but Mohammad Reza Shah's response was a statement in August 1975 that denounced the "extortionate pricing" in the bazaar. In a nationally televised interview, he spoke out against "the exploitation of man by man" and vowed to bring the inflation rate down to zero. The government rolled back the prices of thousands of goods and services to 1974 levels and began importing massive amounts of basic goods and foodstuffs to undercut the bazaar.[77] The monarch also introduced the 14th Principle of the White Revolution: the "Price Stabilization and Campaign against Profiteering." Through this, the government launched a "merciless crusade against profiteers, hoarders, and unscrupulous capitalists."[78] Small shopkeepers were especially targeted, but fines were also levied against wealthy bazaaris to publicize the campaign.[79]

The antiprofiteering campaign was carried out by the newly formed *Rastakhiz* political party. While the state had been operating in a two-party system consisting of *Mardom* and *Iran Novin*, Mohammad Reza Shah liquidated the two and established Rastakhiz in March 1975. Rastakhiz incorporated both capitalistic and socialistic elements to appeal to a broad spectrum of the population,[80] and membership in the party was compulsory for Iranians. Rastakhiz party branches were established in the bazaar, and merchants were forced to donate to the party.[81] That same summer, Mohammad Reza Shah sought to control the guilds. In June, the government liquidated the bazaar's independent guild unions and replaced them with a system that was more conducive to the monarchy's interests. The guild unions were reformed into a new guild system operating out of a Chamber of Commerce, and often led by a non-bazaari.[82] Although governmental pressure over the guilds had been increasingly exercised since the days of Reza Shah,[83] the new state-sponsored guilds were supervised by Mohammad Reza Shah's secret service agency, SAVAK (*Sazman-e Etela'at va Amniyat-e Keshvar*).[84] The government also established a guilds court system—the "watchdog" for Rastakhiz in the bazaar—that would arbitrate charges, levy fines, and enforce sentences of jail terms and exile to bazaaris found guilty of profiteering.[85]

Rastakhiz recruited about ten thousand students to serve as inspectors and, throughout a two-year period, more than twenty-thousand party members inspected prices in the bazaar. Some shops that were found guilty of overcharging were ransacked.[86] During this time, the government levied about 250,000 fines, banned 23,000 business owners from their home towns, sentenced 8,000 to prison terms, and charged another 180,000 with price violations.[87] Misagh Parsa explains how the campaign humiliated bazaaris found guilty of profiteering: "A large banner was hung from the doorway of the shop proclaiming that the store would remain closed for a period of time . . . names and localities of those shopkeepers who were arrested or fined were published daily in the national newspaper."[88] Even though the campaign ultimately had little effect on inflation, bazaaris were severely impacted.[89] As further indication of the stark divide between private-sector elites and nonelites at the time, not a single person from the Chamber of Commerce or the bazaar's parliamentary representatives protested about the campaign.[90]

Other policies were adopted that challenged the ability of the bazaar to conduct business. These ranged from arbitrary taxation to destroying sections of the physical bazaar area. In 1976–1977, Mohammad Reza Shah's government destroyed a large part of the main bazaar in Mashhad, the second-largest city in Iran at the time.[91] Plans were discussed for building an eight-lane highway that would run straight through the Tehran bazaar. The bazaar would be destroyed and in its place a Covent Garden–style market would be constructed.[92] As a point of comparison, Reza Shah leveled small sections of bazaars in Kermanshah, Shiraz, and Yazd, but this was done for the purpose of making new roads. And at least in the case of Yazd, destruction of the bazaar was minimized.[93]

Although the bazaar was subjected to intense state repression and unable to shape the contours of policy making at the time (or more likely because of this), they ended up playing a large part in the revolutionary movement that would bring about the downfall of Mohammad Reza Shah. They could not alter policy making, but they could join forces with individuals who were seeking to bring down the monarch. Their role in the revolutionary movement was defined by orchestrating strikes and shutdowns, financing opposition groups, supporting leaders such as Ayatollah Khomeini, and conducting other activities such as pamphleteering and circulating information to the public. Many bazaaris who took part in the revolution cited "lack of access to state institutions and resources" as a main grievance with the government.[94] Mohammad Reza Shah had fostered a business environment in which a "market mechanism" was lacking, and consequently they turned against the state that was the source of their economic injustice.[95] As the preceding analysis reveals, this lack of access to state institutions, resources, and a market mechanism was the consequence of policy making that emerged because of the oil windfall, rent seeking, and Mohammad Reza Shah's reliance on rents in order to garner political support.

To date, the research on Iran has not focused on the oil windfall as a way to explain Mohammad Reza Shah's policy making toward the bazaar. In one sense, it is very tempting to argue that political instability and fear of regime breakdown is probably what caused him to clamp down on the group. But this is post-hoc theorizing, and we cannot assume there was high political instability during his rule just because a revolution ultimately transpired. Data indicate that there was only a moderate degree of political instability from 1953 to 1978. During these years, Iran saw more political assassinations and purges than the global average, but only slightly more riots and antigovernment demonstrations, and fewer government crises and general strikes.[96] Only one period of high instability stands out—following a massacre of protestors on June 5, 1963. After this event, the government used surveillance and repression in order to prevent the rise of antistate movements. Overall, we have little reason to think that the contours of policy making toward the bazaar are explained by Mohammad Reza Shah's fear of regime breakdown. If he was really worried about the stability of his rule would it not have been easier to simply buy off the bazaar instead? The oil wealth certainly made this possible.

Alternatively, Mohammad Reza Shah's policy making toward the bazaar is explained as a classic "modernization versus traditionalism" scenario.[97] He even wrote in his memoirs that "moving against the bazaar" was a necessary part of "my drive for modernization."[98] While there is good reason to be skeptical of his altruistic intentions, the logic can be turned around to suggest that he was forced to modernize because he could not rally political support from the traditional elements of society, such as the bazaar and the clerics who eventually gained power.[99] Furthermore, the modernization versus traditionalism argument falls apart once the analysis is extended to postrevolutionary Iran. Contrary to what the events of 1979 might predict, we still continue to see rent seeking, preferential treatment for a new group of elites, and a business environment that inhibits the activities of domestic entrepreneurs and small-sized firms. Despite their sacrifices for the revolutionary movement, the bazaar does not fare much better in the Islamic Republic.

Finally, it may be argued that Mohammad Reza Shah's policy decisions were less a reflection of the impact of oil and more an example of policies designed to achieve large-scale industrialization during the White Revolution. Indeed, it may seem difficult to separate out the effects of oil from the effects of the White Revolution because they both emerged within the span of a single decade. During the period between the oil windfall and the introduction of the White Revolution—roughly 1955 to 1962—the country experienced a brief boom in investment but then a balance of payments crisis and increasing political uncertainty for the monarch. But there is also evidence that the dynamics that came to define the White Revolution era—that is, rent seeking from elites and preferential treatment from the state—manifested *prior* to the introduction of the White Revolution in 1963. Consider the following statement by Karshenas: "From the late 1950s a new breed of industrialists with close ties to the state and to the new financial institutions flourished, who increasingly relied on cheap loans from the industrial banks to invest in new industries."[100] The White Revolution may have highlighted rent seeking from elites and gave Mohammad Reza Shah a chance to adopt preferential policies to benefit this group, but these dynamics emerged earlier as a result of the oil windfall.

4.4 THE OIL ERA: POLICY MAKING AND THE PRIVATE SECTOR IN THE ISLAMIC REPUBLIC

The 1979 revolution removed Mohammad Reza Shah from power and ended centuries of monarchical rule in Iran. Although the event was understood by many as a victory over the monarchy's policy failures, it led to considerable uncertainty and anxiety about the direction of the country's economic planning. The general perception at the time was that, even though Iran's economy experienced considerable growth during the 1960s and 1970s, all of the benefits went to elites. Among the Islamic clergy who assumed a leadership role after the revolution, most advocated antidependency, social welfare, and greater redistribution of oil revenues. However, a heated debate began about the role of the private sector. In spite of the bazaar's numerous and visible contributions to the revolutionary movement, a general atmosphere of anticapitalism permeated the early days of the Islamic Republic.[101] For certain, this reaction was in part fueled by the excesses of the Shah-era elites and the rent-seeking behavior that funneled oil revenues into the hands of the few.

Although more than thirty-five years have passed since the revolution, many would argue that a certain element of these feelings still persist in Iran. The word for "capitalist" in Persian (*sarmayehdar*) carries a pejorative connotation and many Iranian entrepreneurs today prefer to call themselves *karafarin*, or "creator of work."[102] The enduring negative characterization of private-sector actors serves a political purpose, however, because it helps justify an approach to policy making that preferences the role of *bonyads* (quasipublic foundations) and related state-supported organizations in the economy. Yet, as I illustrate in the following text, the rise of *bonyads* is less a story about ideology and more an outcome associated with the oil windfall. Although some portray the rise of *bonyads* as a unique development in Iran,[103] I highlight the similarities between their rise to power and the rise of private-sector elites under the rule of Mohammad Reza Shah. First, I provide an overview of policy making toward the private sector in the postrevolution era before moving on to the evidence for the microfoundations of the theory.

After a power struggle following the events of 1979, Ayatollah Khomeini and the pro-statist camps ultimately gained leadership of Iran's government. They focused on enhancing the role of the state in the economy, but they also acknowledged that the private sector had a part to play. This was due in some measure to the key role that the bazaar played in the revolution and a broader recognition of a close bond between Muslim clerics and the bazaar during watershed moments in Iran's history. Khomeini repeatedly stated that commerce and merchant activities were important to the health of Iran's economy. He pointed to the bazaar's financial support for building mosques and uniting with the clergy during the tobacco concession crisis as well as other instances of collective action against the now-ousted monarchy. Ayatollah Khamenei similarly said that Islam has always supported the bazaar and commerce, and in reality it is the socialists who link business to "theft, corruption, and exploitation."[104] There are echoes of these sentiments today. Ali Fazli, Deputy Commander of Iran's Basij,[105] told the media in February 2013: "No one has denied the role of traders in the victory of the Islamic revolution. . . . Traders are the driving strength of the revolution and when any problem emerges the traders are the first to stand in line."[106]

Consequently, even when the newly established government expropriated firms, factories, and wealth from the former private-sector elites, policy makers "carefully avoided challenging the concept of private property."[107] Instead, they justified their actions as punishment against enemies of the revolution. Nevertheless, the government's actions sent mixed signals to private-sector actors about their future role in Iran's economy, and some of these mixed signals appear today. Rhetorical support for the private sector is combined with warnings against exploitation, profiteering, and neglecting the welfare of the people.[108]

The rise of Khomeini and the presidential administration of Khamenei (1981–1989) led to an economic program that increased state intervention in the economy. For the moment, the private sector was pushed to the side. The eight-year war with Iraq began in 1980, and the government put itself in control of markets and all major industries from oil to large-scale manufacturing, banking, transportation, and communications. By the late 1980s and after the war ended, it was estimated that the government controlled 60 to 70 percent of the domestic economy and 90 percent of foreign trade. Although Mohammad Reza Shah's government exerted a high degree of control over the economy, some analysts observed that the postrevolutionary government intervened at a greater degree.[109] The central role of the state combined with the chaos of war and uncertainty about the private sector's role in the economy affected private investment, which dropped 5 percent from 1980–1981 to 1989–1990.[110] State expropriations and confiscation of private capital increasingly led to a "virtual withdrawal" of the private sector from industry.[111] Erratic levels of private investment lasted throughout the late 1990s.[112]

After the war with Iraq came to an end, Ayatollah Rafsanjani was elected president (1989–1997) and there were louder calls for a more active role for the private sector in the economy. Rafsanjani is often viewed as the most "pro-private sector" president that Iran has had to date, and upon his election the business community was hopeful that he would take steps toward economic liberalization. Given that a brutally destructive war with Iraq had recently concluded, Rafsanjani was responsible for rebuilding the economy, a task in which the private sector could be called upon to play a fundamental role. But in spite of his reputation and his "consistent attempt to invigorate the private sector,"[113] few reforms were adopted while he was in office. Although oil prices were also low during this period, Rafsanjani was unable to shift Iran's momentum toward business environment reforms.

His successor, Seyed Mohammad Khatami (1997–2005), similarly did not veer from the path. In fact, Khatami was largely responsible for turning the public's focus away from the economy and toward political and social issues. Some analysts remarked that he appeared bored with economic issues.[114] On the eve of the start of his second term, it seemed that "Khatami is not a president for economic reform."[115] That being said, a few developments took place under his leadership that positively affected the domestic private sector. One was in banking. After the revolution and during the war with Iraq, the government nationalized banks and introduced Islamic banking. During Khatami's tenure, the private sector could be granted licenses to open banks, insurance companies, and credit institutes. Steps were made toward the promotion of investment.[116] This led to increasing levels of credit and investment in the private sector. In addition, the modest political and social freedoms granted under Khatami may be one reason why some reported "a genuine growth in private sector confidence in economic activity."[117] Nevertheless, many observed that Khatami's administration gave the same unclear signals toward the private sector as that of his predecessors. Last ditch efforts to propose economic reforms during his final years in office were blocked by his political opponents.[118]

The presidency of Mahmoud Ahmadinejad (2005–2013) was similar to Khatami's in that economic issues were less at the forefront than political and social issues. Under Ahmadinejad, some key developments took place, most importantly the increased role of the *bonyads* and the increased economic activities of *Sepah-e Pasdaran-e Enqelab-e Islami* (the Islamic Revolutionary Guards Corps, or IRGC). Credit lending decreased under Ahmadinejad. Although net domestic credit was increasing every year since the end of the Iran–Iraq war (albeit not as a percentage of GDP), it dropped 36 percent in 2010. From Khatami's first year in office until his final year in 2005, net domestic credit to the private sector was increasing year-on-year and overall nearly 300 percent throughout his two terms. Credit grew at a similar pace during Ahmadinejad's first term, but fell sharply in 2010.[119] As credit declined, there were increasing signs of credit deferments and defaults from the private sector.[120] Overall, and in spite of any rhetoric, Ahmadinejad's administration seemed reluctant to embrace the idea of boosting the private sector's role in the economy.[121]

Ahmadinejad's time in office coincided with greater dialogue about the privatization of state-owned enterprises. But even here, as Kevan Harris argues, the government appears to have more of a "pseudo-privatization" in mind.[122] Prior to Ahmadinejad's first term, the Supreme Leader Khamenei passed a decree to begin large-scale privatization of many nonoil sectors. Implementing this decree has been a slow process, and there is criticism targeted toward the government about which actors are actually benefiting from the sale of state enterprises. A number of observers point out that most enterprises are being sold to rent-seeking *bonyads*, the IRGC, or individuals with close personal ties to policy makers. Next, I examine the government's relationship with these actors in detail.

4.5 THE OIL ERA: RENT SEEKING AND SUPPLY-SIDE FORCES IN THE ISLAMIC REPUBLIC

The role of *bonyads* adds an important nuance to standard notions about the mix of "public versus private" in the Iranian economy. In some ways, the rise of *bonyads* after the revolution may seem paradoxical, given the overall statist approach to the economy and the fact that many *bonyads* are massive conglomerates who dominate the nonoil economy and possess considerable autonomy from the government. I argue that a sense of paradox is averted if *bonyads* are understood as playing a similar political role as the former private-sector elites during the rule of Mohammad Reza Shah. Much like the former private-sector elites, *bonyads* engage in highly successful rent seeking and are granted preferential status in many areas of the economy in which they operate. Major investments are concentrated in the hands of a few *bonyads*, just as major investments were concentrated in the hands of a small group of private-sector elites. *Bonyads* tend to be staunch political supporters of the government, again much as the Shah-era elites were. These facts notwithstanding, one crucial difference between modern-day *bonyads* and the Shah-era elites is that the former provide a range of welfare activities alongside their economic activities.[123]

Bonyads were not invented in 1979. They existed during the rule of Mohammad Reza Shah where they served a dual purpose: on the one hand, a charitable organization that helps provide welfare services for the needy; on the other, a tax-free conduit for funneling oil revenues toward the government's investments. But whereas Mohammad Reza Shah and the royal family maintained complete control over the activities and decision-making aspects of these foundations, modern-day *bonyads* operate independently from the government while still remaining highly dependent on the rents and subsidies it provides. *Bonyads* receive as much as 58 percent of the government's budget, but the elected branches of government "do not have any authority to monitor their performance."[124]

The rise of *bonyads* began shortly after the revolution as Mohammad Reza Shah and his family fled the country and were followed by many of the private-sector elites who prospered during his rule. These elites took whatever assets they could with them and abandoned their firms and factories. In some cases, workers organized takeovers of the abandoned firms and factories. For the most part, the government ended up confiscating thousands of firms that spanned the industrial, agricultural, and trade sectors.[125] Firms were placed into four groups according to the "Law for Protection and Development of Iranian Industries," with the government gaining control of the first three groups. These were the heavy and strategic industries, firms whose owners had close links to the monarch, and bankrupt and debtor firms. When it was determined that a firm had a "legitimate" owner it did not come under government control.[126]

Many of the confiscated firms and holdings were given to newly established revolutionary *bonyads*, in particular *Bonyad-e Mostazafan* (Foundation for the Oppressed) which acquired more than 110 industrial units. Other *bonyads* were formed such as *Bonyad-e Shahid* (Martyr's Foundation), *Bonyad-e Panzdah-e Khordad* (15th of Khordad Foundation), *Jehad-e-Sazandegi* (Reconstruction Crusade), and *Komiteh-ye Emdad-e Emam* (Imam's Assistance Committee). In conjunction with managing their economic holdings, each *bonyad* was tasked with a welfare activity. For instance, *Komiteh-ye Emdad-e Emam* was tasked with providing assistance to the poor, particularly in rural areas. *Bonyad-e Shahid* was tasked with providing families of individuals who were killed during service to the government, such as in the Iran–Iraq war. *Bonyad-e Panzdah-e Khordad* was tasked with providing support to widows, orphans, and war victims.[127]

Alongside their inheritance of a wealth of capital and investments, the government gave *bonyads* generous bank credits, free loans, privileged access to fuel, and reduced-rate foreign exchange. They were granted exemption from the corporate tax, received a bevy of government contracts, were given monopoly status in many sectors, and successfully limited license granting in order to prevent competition.[128] In sum, *bonyads* are "unregulated and unclear para-governmental organizations . . . [they are] conglomerates which benefit substantially from both implicit and explicit government subsidies."[129] This has resulted in the rapid growth of *bonyads'* activities since the revolution, and by the late 1990s they gained control of up to 40 percent of Iran's nonoil economy. The dominant role of *bonyads* in nonoil sectors is facilitated by their ability to tightly protect their rents and autonomous status. They are "neither subject to the forces of the market, nor to the budgetary and accounting scrutiny of the government."[130] The government does not have the ability to monitor their expenses, there are no shareholders, and their legal status is vague; *bonyads* are "effectively a law unto themselves."[131] The role of *bonyads* in

Iran's economy has had a major impact on the contours of policy making. Their "economic and political power has prevented any discussion of making the slightest limitation in the expanding domain of their activities."[132]

As the preceding descriptions suggest, *bonyads* are distinct from the public sector but possess a considerable array of advantages over any other private-sector actor in the country. They are dependent on the government for their rents and protected status, yet they exercise considerable autonomy from the government. The largest *bonyad* is *Bonyad-e Mostazafan*, the second-largest commercial enterprise in Iran (after the national oil company) and one of the largest holding companies in the Middle East. *Bonyad-e Mostazafan* is described as "institutionally and legally, the most obscure of the foundations."[133] Since the early 2000s, *Bonyad-e Mostazafan*'s holdings include four hundred companies, which alone are estimated to control more than 25 percent of Iran's nonoil economy in sectors from textiles to petroleum services to soft drinks.[134] *Bonyad-e Mostazafan* directly employs more than seventy thousand people, and indirectly perhaps four hundred thousand people.[135]

The rapid increase in the size and relevance of *bonyads* in Iran's economy since 1979 can be witnessed in a display of employment data in Figure 4.4. Despite their vast array of economic activities, *bonyads* are classified by the government as belonging to the community and social services sector. In the years from 1976 to 1991, employment in the community and social services sector eventually surpassed the agricultural sector (not displayed), long the sector in which most Iranians worked. Although employment in

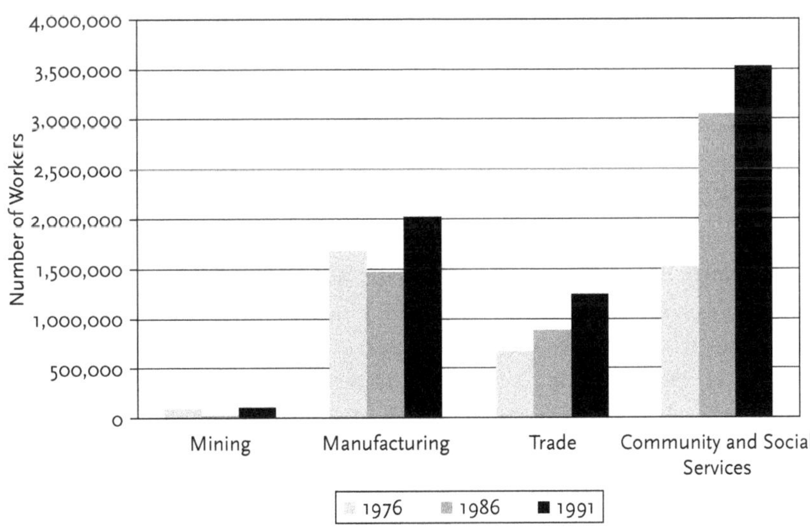

Figure 4.4 Total Employment in Iran by Sector and Year

agriculture increased 7 percent from 1976 to 1991, the community and social services sector increased more than 131 percent. By comparison, manufacturing increased slightly, and remained the same as a share of the population. Employment increased somewhat more in trade.

Bonyads are an important source of political support for Iran's clerical leaders. *Bonyads* are "absolutely loyal to the supreme leader and to local religious leaders who act as representatives of the supreme leader in the capital."[136] As Ali Saeidi explains, "[*bonyads*] were established in order to assist institutionalization of the ideology of the ruling class . . . when the revolutionary forces could not trust the old regime's bureaucratic apparatus."[137] In one sense, given that redistribution and welfare are central themes of the postrevolutionary government, *bonyads* play an important role as substitutes for state capacity in social service provision—and thereby help boost the legitimacy of the state. In another sense, because *bonyads* are dependent on state-granted rents they remain active supporters of its leaders while being protective of the privileged relationship they have. Unsurprisingly, *bonyads* are "intrinsically opposed to any opening of the economy to 'outsiders.'"[138]

Alongside *bonyads*, the IRGC is another group to successfully reap benefits from the conversion of the oil windfall into rents. Although their main task is serving as the country's security apparatus, the IRGC has extended their activities into the economic and infrastructural realms. They have received many government contracts, such as those for dam construction, telecommunications, electronics, and defense-related industries.[139] During the mid-2000s and after the election of Ahmadinejad, the government began granting billions of dollars in contracts to the IRGC. The IRGC's engineering and construction unit, *Khatam al Anbia* (GHORB), has leveraged itself as a primary beneficiary of these contracts through its close ties with the government. The government granted GHORB contracts without participation of any other bidders and funded the projects from the country's Oil Stabilization Fund (OSF). Established under Khatami in 2000, the OSF's purpose was to save and invest Iran's oil revenues in order to forestall against fluctuating oil prices.[140] The allocation of OSF funds to the IRGC led some politicians to demand an explanation from Ahmadinejad regarding "whether allocating such funds . . . to the [IRGC] was legal and whether participation of . . . [GHORB] in economic projects was part of its mandate."[141] It is worth noting that the OSF was arguably the only policy tool that could rein in the state's spending of oil revenues. But many reported on the depletion and subversion of the OSF during Ahmadinejad's presidency, and in 2011 the Iranian media stated that the OSF's balance was zero.[142] A report from Revenue Watch Institute estimated that from

2006 to 2011 the government withdrew more than 150 billion US dollars from the OSF "without clear economic justification."[143]

Bonyads and the IRGC have been the main beneficiaries of the privatization of state-owned enterprises. In 2004, the government decided to increase privatization by amending Article 44 of the Constitution and established a goal of privatizing 80 percent of the shares of state-owned enterprises.[144] Since then, privatization has moved slowly and benefited those with close ties to the government, such as *bonyads*. The slow movement of privatization is sometimes justified as a need to proceed cautiously in moving the economy away from an oil-dependent state. In 2013, a former head of Iran's Privatization Organization, Seyed Ahmad Mir-Motahari, noted the difficulty of transitioning away from oil dependence, the need to not make such a decision "hastily," and the need for a transition period.[145] Others in government, such as Parviz Davoodi, vice president during Ahmadinejad's first term, called for increasing the pace of privatization.[146] Ahmadinejad praised Article 44 near the end of his second term, describing it as "one of the most important, most fundamental, and most transformative laws that has been adopted in the country until now."[147]

There are a few reasons why privatization has not proceeded steadily or smoothly in Iran. Some argue that the mind-set for privatization is not yet present in the government. Others point out that most of the units privatized face enormous losses.[148] Another former head of Iran's Privatization Organization, Mohammed Rahim Ahmadvand, stated that the barriers to privatization emerge from within the government and there is a lack of cooperation among governmental branches on implementation of Article 44.[149] Parliamentarian Hamid Fouladgar made the more critical point, which others have noted, that the government's assets are not really being privatized but simply circulating within the government.[150] The Iranian media similarly underscores the surging power of *bonyads* during the privatization process, with whom the private sector cannot realistically compete against.[151]

The government's preferential treatment of rent-seeking *bonyads* and the IRGC is contrasted with excessive and complex restrictions on the majority of entrepreneurs and owners of modest capital. This is in spite of the fact that small- and medium-sized firms continue to play a vital role in Iran's economy. Even as *bonyads* and other state-sponsored units expand their economic activities, SMEs still account for as much as 94 percent of all firms in Iran.[152] As I discuss next, the Iranian government has adopted policies in the postrevolution era that place many restrictions on the growth of SMEs in key nonoil sectors, such as manufacturing. Additionally, the government has weakened SMEs and nonelites on political terms, and thus they have been unable to influence the contours of policy making toward the business environment.

4.6 THE OIL ERA: POLICY MAKING, NONELITES, AND DEMAND-SIDE FORCES IN THE ISLAMIC REPUBLIC

During the postrevolution era, the Iranian government has imposed a range of regulatory requirements, tax burdens, and restrictions on domestic entrepreneurs and privately-owned SMEs. The regulatory bureaucracy in postrevolutionary Iran is often described as an intricately layered system of controls and monitoring of the private sector. As two examples, into the late 1990s, import licenses were overseen by more than thirty different state entities and sixteen different state entities dealt with price controls. Top-level governmental branches impose an array of rules and regulations that lower-level bureaucrats are supposed to implement, but it can be difficult to keep up with the sheer number of rules handed down. In 1997, the customs administration complained that they had to enforce eight hundred directives from the Central Bank.[153] Regulatory policies are known for being complicated and changing rapidly.[154] Hamid Zangeneh holds that, "At all levels, bureaucrats have been unable or unwilling to make decisions that can stand the test of the day."[155]

There is evidence that small-sized firms have faced disproportionately higher tax burdens than larger firms in the postrevolution era. Data indicate that indirect taxation of the manufacturing sector jumped from 1998–1999 to 2005–2006, even as Iran's oil income was increasing over this period.[156] Table 4.3 displays the average annual bill of indirect taxes and charges for firms (measured per worker in millions of rials) in four manufacturing sectors: textiles, wood products, paper, and furniture.[157] In 1998–1999, the average tax bill per worker across all sizes of firms was

Table 4.3 AVERAGE ANNUAL INDIRECT TAX BILL PER WORKER FOR IRAN'S MANUFACTURING FIRMS, BY SIZE AND TYPE OF FIRM (IN MILLIONS OF RIALS)

	Large size: 100 and + workers		*Medium size: 50 to 99 workers*		*Small size: 10 to 49 workers*	
	1998–1999	2005–2006	1998–1999	2005–2006	1998–1999	2005–2006
Textiles	0.204	0.524	0.428	0.56	0.287	0.633
Wood Products	0.395	0.272	0.11	0.243	0.135	0.34
Paper Products	0.178	0.6	0.555	0.475	0.466	0.761
Furniture	1.734	0.277	0.353	0.432	0.203	0.578
average	**0.628**	**0.418**	**0.362**	**0.428**	**0.273**	**0.578**

0.31 million rials in textiles, 0.21 million rials in wood products, 0.4 million rials in paper, and 0.76 million rials in furniture. But there is variation within sectors according to the size of the firm. In 1998–1999, the smallest firms had a moderate tax bill in textiles and paper, and a low tax bill in wood products and furniture. Medium-sized firms in the textile and paper sectors had the highest tax bill, and large-sized firms in the wood products and furniture sectors had the highest tax bill. However, by 2005–2006, the *smallest* firms across all four sectors paid the *highest* bill for indirect taxes and charges per worker.

Data from the World Bank reveal that Iran imposes relatively higher taxes and mandatory contributions on medium-sized companies.[158] As of December 2011, the profit tax and total tax on profits were higher than the global median. The total estimate of tax hours spent per year and the percentage of labor tax and contributions were both significantly higher than the global median (above the 75th percentile). Although Iran had fewer tax procedures (twenty to the global median of twenty-eight) and fewer "other taxes," it was ranked 129th overall out of 185 countries with respect to paying taxes for a medium-sized firm. To be sure, Iran's imposition of higher tax-related burdens on these types of firms runs contrary to what the rentier state literature suggests about oil-rich countries, which are assumed to be relieved of the need to tax. But Iran is not alone in its pursuit of tax revenue from the private sector, as other oil-rich countries today are similarly seeking out this source of governmental revenue.[159]

Beyond the formal regulatory policies and procedures, there are reports of "hidden barriers" to the activities of entrepreneurs and firms in Iran. Public officials have some discretion about determining who can legally open a business and "individual government officials could and would interpret the law to suit the person and not the situation."[160] It is also reported that firms often need to engage in bribery. According to Hossein Askari, "To get anything done in Iran, from getting a building license or clearing goods through customs, may well require a payment of some kind."[161] The practice of bribery also stems from the fact that regulatory rules and policies are not evenly applied throughout the private sector. A multilayered system of bureaucratic oversight provides multiple entry points for individuals to use bribes as a way to circumvent regulation.[162]

Credit to the private sector as a percentage of Iran's GDP has steadily decreased since 1988. Credit allocation continues to be highly personalized and only *bonyads* and rent-seeking elites have regular access to the low-interest credit that is facilitated by the oil windfall.[163] Access to credit is routinely cited as a main obstacle to entrepreneurs, and banks exhibit a reluctance to extend credit to small firms—forcing them to go to the bazaar where, as during the time of Mohammad Reza Shah, interest rates tend to be higher.[164] The inability of small firms to access credit has a direct effect on their ability to stay competitive and innovate. In a survey of fifty SMEs located in Iran's Sistan and Baluchestan region, owners ranked "cost of finance" and "availability of finance" among the top three reasons why they encounter difficulties in innovation.[165]

The regulatory, taxation, and credit-related burdens imposed on SMEs extend to those located in the bazaar, who have faced heavy regulation since the revolution. This may come as some surprise given what we know about the role of the bazaar in the revolutionary movement, as well as their lengthy ties to many of the clerics who eventually became leaders in the Islamic Republic. We might have expected the postrevolutionary government to reward bazaaris by allowing them to continue their activities relatively unfettered or buy off the group as a means of bolstering support for the new government. The fact that the government instead relied on regulation provides additional support for the theory that an oil windfall leads to a restrictive business environment for nonelites—irrespective of the type of government or even if the nonelites were frequent allies of the new leaders. According to a detailed study by Arang Keshavarzian, the government's regulation of the bazaar after the revolution is based on a desire to "depoliticize" the group, which weakens their ability to shape the contours of policy making. Although some well-connected bazaaris were rewarded for their sacrifices during the revolution, these individuals were then used by the government to monitor the activities of others.[166]

Policy makers in Iran sometimes pander to the bazaar, but there are also examples of rhetoric that reflect a more uneasy government–bazaar relationship. During the late 1980s and after the Iran–Iraq war, some of the influential members of the Islamic clergy portrayed the bazaar as exploitative capitalists and criticized pro-private-sector clerics for being "more pro-free enterprise than even Western capitalists."[167] In 2011, Ahmadinejad highlighted the important role the bazaar plays in Iranian society and economy, but he also added that there are some bazaaris who "give the bazaar a bad name." He said that the bazaar has the ability to self-regulate, but encouraged members to cooperate with the government on issues of supervision and inspection.[168]

Hostile confrontations between the government and the bazaar has emerged in recent years, although they are less intense than what happened during Mohammad Reza Shah's rule. For instance, bazaars went on strike in October 2008 and July 2010 to protest tax increases. During the latter strike, the riot police and paramilitary entered the Tehran bazaar and one bazaari was killed.[169] In October 2012, bazaars in Tehran and Esfahan closed in protest after the value of the rial fell 40 percent in one week. The government blamed the currency fluctuations on "unscrupulous" members of the bazaar.[170] The riot police arrested at least sixteen currency traders during the shutdown,[171] and many shop owners closed their shops out of fear that their property would be destroyed by government-supported enforcers.[172] A police chief told the media that bazaar units can legally close for up to fifteen days, but closing with the purpose of disrupting the economic system is a "crime to be dealt with."[173] In August 2014, the equipment and tools shop-owners' guild and ironmongers in the Shahabad Bazaar in Tehran went on strike to protest what they viewed as excessive taxation.

Policy makers had mixed responses to the October 2012 crisis in the bazaar. One parliamentarian, Nader Ghazipour, saw the government's intervention in the currency market as the main reason for the turbulence in Iran's currency.[174] Habibollah Asgaroladi, secretary general of the Islamic Coalition Party (a party originally founded as the political and security wing of the bazaar), viewed the event as further evidence that the government needs to attend to the needs of the bazaar to stop manipulative business activities from flourishing.[175] The head of the National Union of Guilds placed the blame on the government and economic ministers. He said that their decisions produced "destructive" outcomes as they continue to believe that trade "has no place in the economy."[176]

The political weakness of the bazaar is coupled with ineffective formal business associations. Since the revolution, the country's business associations have been described as either weak, lacking autonomy, or in conflict with the government. In 2009, Hossein Noghrehkar-Shirazi, deputy oil minister and government representative on the board of the Tehran Chamber of Commerce, observed: "I have been present in chambers of commerce for 28 years and, so far, there has always been a state of opposition (between chambers and the government)."[177] Iran's Chamber of Commerce, Industries and Mines (ICCIM) is often referred to as a "semigovernmental organization," and many stress the need for more autonomy. In 2011, Pedram Soltani, vice president of ICCIM, argued that its role should be to push the government to protect the rights of the private sector. According to Soltani, it should employ tactics that range from "criticism and protest" to "association and interaction."[178] In recent years, the ICCIM has criticized the government for its handling of privatization, preferential treatment of *bonyads*, and with respect to its promised benefits from the country's subsidy reform program. The government planned to begin the second phase of subsidy reforms in June 2012 (they were temporarily suspended), which would redistribute benefits to industry.

Yet, Mohammad Nahavandian, then president of ICCIM, pointed out in January 2013 that the government still had not paid 30 percent of its commitments to industry during the first phase of subsidy reforms.[179] As during the time of Mohammad Reza Shah, business associations remain weak in their ability to encourage the government to implement policies and practices that would benefit the private sector as a whole.

The government's heavy regulation of the bazaar and nonelites has had measurable effects on the growth of the private sector since the revolution. A decline in private entrepreneurship in the manufacturing sector is evident by looking at the available data on own account workers.[180] Own account workers are self-employed individuals who do not have any employees. In the context of manufacturing, they tend to be artisans or run their own workshops. The contribution that own account work makes to private-sector development is minimal, and they have no broader impact on unemployment. Own account work is generally viewed as a subsistence activity, given that most of these individuals are illiterate and self-employment is often their only option for survival. As Figure 4.5 displays, the own account category had the highest share of illiterate workers in Iran of all employment categories in 1986.

From 1976 to 1991, the percentage of workers in the manufacturing sector classified as own account substantially increased from 18 to 32 percent (see Figure 4.6). Yet, the number of individuals classified as private employers—a more robust indicator of entrepreneurship and private-sector development—grew marginally over this period. From 1986 to 1991, there was effectively no growth in the number of private employers, and this

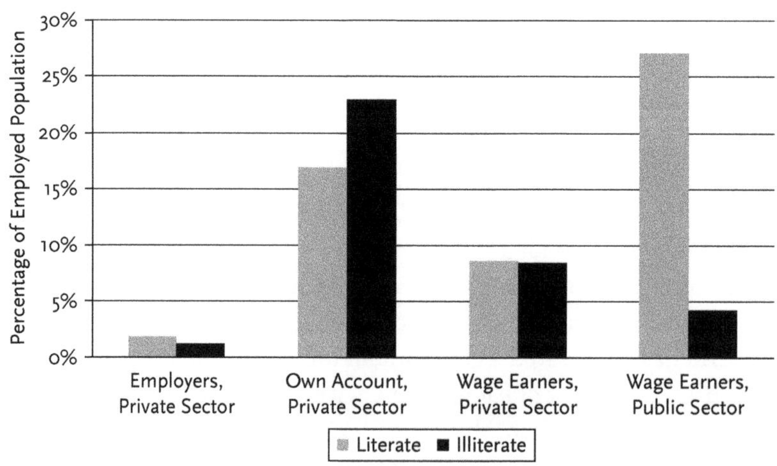

Figure 4.5 Iran's Employed Population According to Literacy Status in 1986

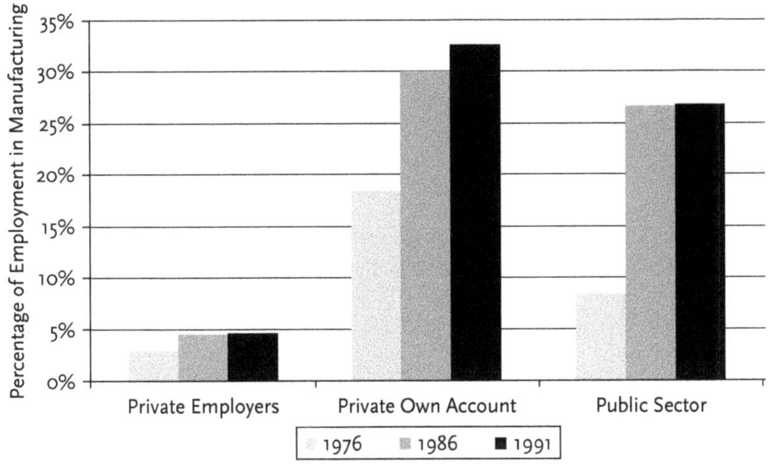

Figure 4.6 Employment in Iran's Manufacturing Sector by Status and Year

group actually shrank as a share of Iran's population. To be sure, the 1976 to 1991 period overlaps with the revolution and the war with Iraq, which may help explain a lack of growth in the private sector. Nevertheless, the 1986 and 1991 data indicate that even during the initial postwar reconstruction era there was no increase in the number of employers but a continued growth in own account workers.[181] This period also coincides with the oil bust, when we should have seen the government making efforts to boost the growth of nonoil sectors. But nearly all employment growth in Iran's manufacturing sector from 1976 to 1991 took place in the public sector or own account activities. Growth in private employers stagnated.

And yet, the manufacturing sector grew rapidly after the revolution. As Figure 4.7 displays, manufacturing slowly grew during the 1960s–1990s and then faster after 2000.[182] However, Figure 4.8 suggests that virtually all growth during the 2000s took place among firms with more than one hundred employees. Given that just 11 percent of the 15,443 privately-owned manufacturing firms in 2006 had more than one hundred employees, we can infer that much of this growth took place in publicly-owned firms or *bonyads*. Credit allocation to the manufacturing sector, in particular, was very low during the period of postwar reconstruction. Credit lending as a percentage of GDP declined during the 1990s, and the manufacturing sector received little credit from the four main specialized banks (see Figure 4.9).[183]

Similar trends are apparent in Iran's trade sector, a sector in which the bazaar has long played a prominent role. Figure 4.10 displays the available employment data and, here again, the number of private employers

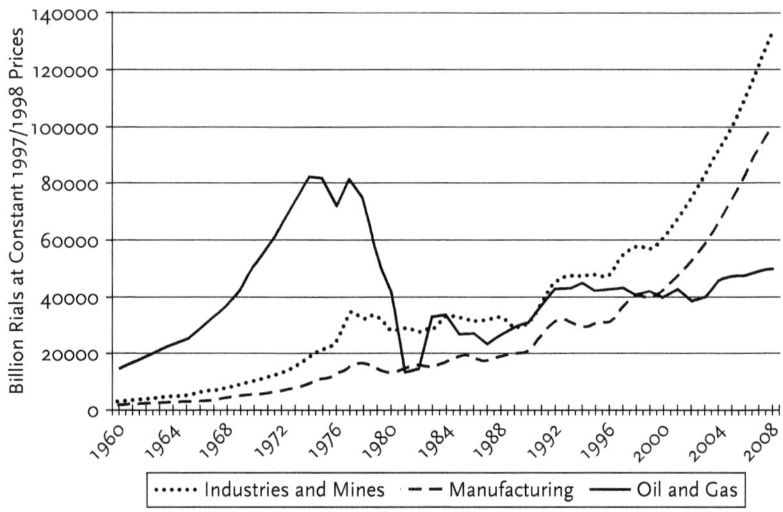

Figure 4.7 Iran's Gross National Product by Activities

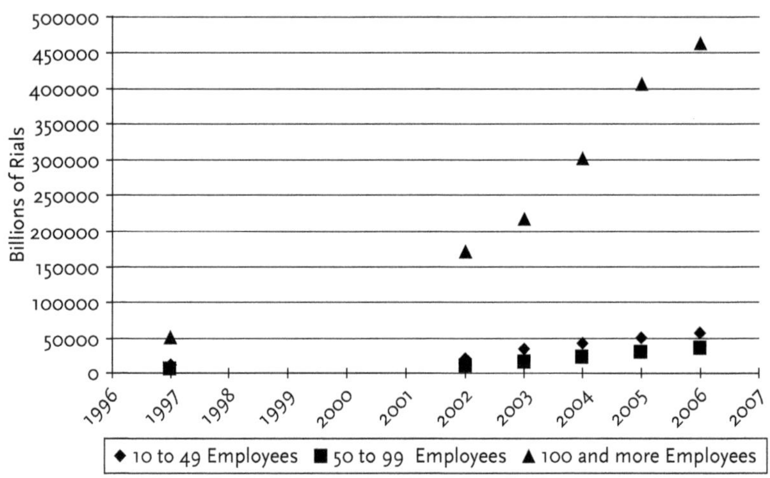

Figure 4.8 Value of Iran's Manufacturing Output by Size of Establishment

declined from 1976 to 1991, as did the share of the population defined as a private employer. The own account classification increases marginally and is the modal category in trade. Most of these individuals would be operating in the physical bazaar. Employment in publicly owned units in the trade sector increases marginally. Thus, in the trade sector we likewise observe negative growth in the robust forms of entrepreneurship (private employers) and a slower but continued dominance by nonrobust forms (own account workers).

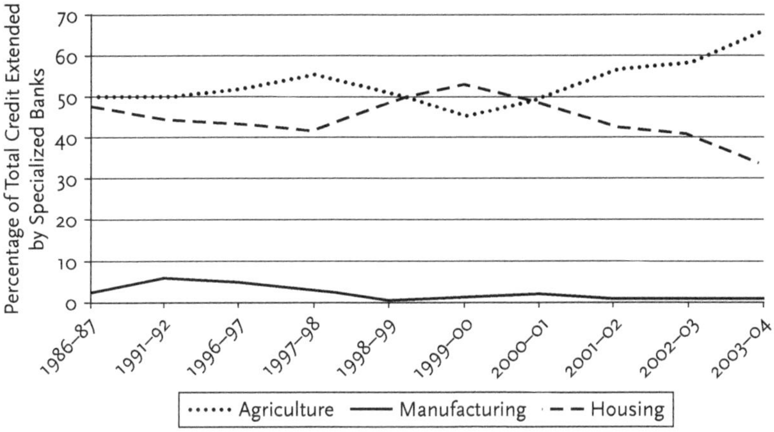

Figure 4.9 Credit Extended by Specialized Banks in Iran to Agriculture, Housing, and Manufacturing

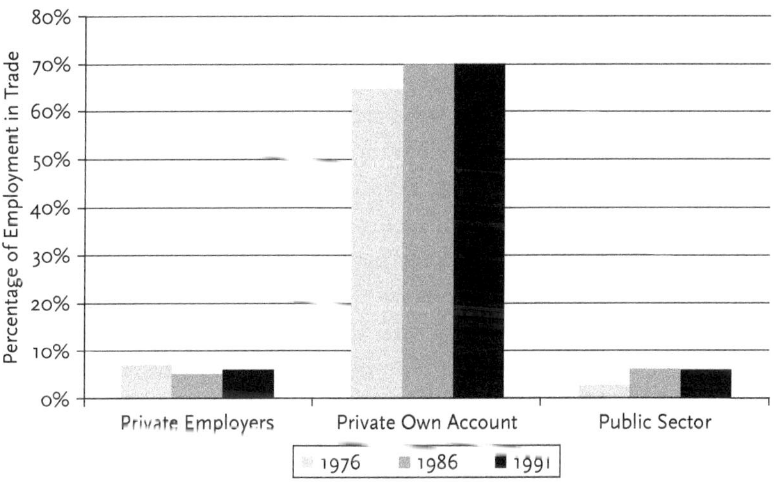

Figure 4.10 Employment in Iran's Trade Sector by Status and Year

More detailed data help reveal trends in the types of trade activity and the increasing role of *bonyads* in trade. After the 1979 revolution, *bonyads* and governmental organs (such as trade units in ministries) increased their role in the trade sector.[184] Large-size *bonyads* have cut into the bazaar's activities in wholesale trade, and to a lesser extent in retail. Consider the fact that, in 1976, just prior to the revolution, there were 17.5 times as many retail units as wholesale units in Iran. But just a decade later this ratio was 5.3, a reflection of the marked increase in the number of *bonyad* and public wholesalers in the country.[185] Figure 4.11 highlights

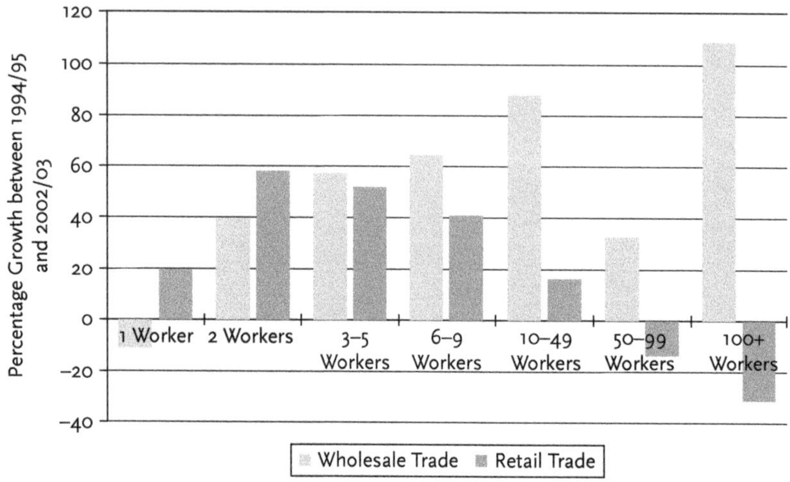

Figure 4.11 Growth in Iran's Wholesale and Retail Trade Units, 1994–1995 to 2002–2003

the extent to which *bonyads* and similar units have increased their activities in wholesale trade over time. From 1994–1995 to 2002–2003, the number of small, bazaar-based wholesale trade units (those with just one worker) decreased while wholesale trade units with one hundred and more workers substantially increased. Yet in retail, the trends were reversed. The largest increase in the retail sector was for firms with just two workers, most of which are located in the bazaar.

As *bonyads* push bazaaris out of the high-growth wholesale sector and relegate them to the low-growth small retail sector, this can have a broader impact on entrepreneurship in small-scale manufacturing. This is because the transition from localized trade to small-scale manufacturing is considered a critical moment in the economic development process.[186] Subnational data at the provincial (*ostan*) level help confirm the link between trade and manufacturing in Iran. They also underscore an important nuance. As Figure 4.12 displays, retail trade has a consistently positive association with private manufacturing as opposed to wholesale trade.[187] The correlation between the number of retail trade units in a province in 2003 and the number of small private manufacturing firms (ten to forty-nine workers) in the same province in 2006 is positive and strong at 0.652. However, the correlation between wholesale trade units in a province in 2003 and small private manufacturing firms in the same province in 2006 is positive but moderate at 0.46. The correlation between retail units in 2003 and the number of new manufacturing licenses granted in 2006 is positive and moderate at 0.365, yet the

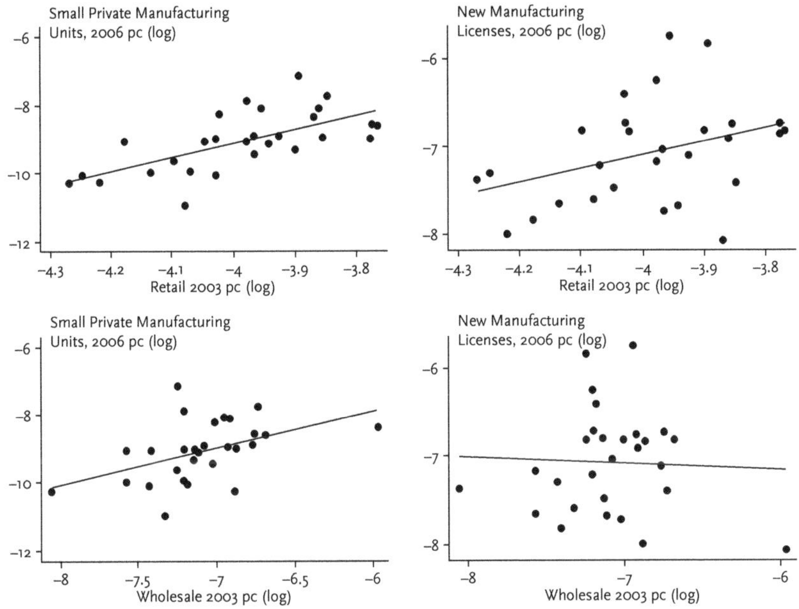

Figure 4.12 Trade and Manufacturing in Iran's Provinces

correlation between wholesale units in 2003 and new manufacturing licenses in 2006 is negative and weak at -0.044. While the data are only suggestive, they may signal that growth in Iran's small private manufacturing is more likely to occur in places where there are more retail trade activities—which we know are mostly driven by private entrepreneurs and not *bonyads*. But retail trade has slowly grown since the revolution compared to wholesale trade. Wholesale trade has grown fast, but it is propelled by *bonyads* and government-sponsored units and has a weaker link with private entrepreneurship.

In spite of the private sector's potential to help reduce unemployment and serve as a possible incubator for innovation, the postrevolutionary government has imposed numerous regulations on entrepreneurs, struggled to provide the institutional foundations to support their activities, and generally created an atmosphere where running a business can be quite difficult.[188] This has negatively affected entrepreneurship and hurt the effectiveness of (the limited number of) government programs that were actually intended to help the private sector after the revolution.[189] Like many governments around the world, Iran's government sends signals that it wants to encourage entrepreneurship, but it adopts policies and practices that undermine this goal. Yet, in order for the government to positively affect private-sector development it needs to seek a holistic approach over time rather than pursue one-off, isolated projects that can be manipulated by individual bureaucrats or wind up on the budget's chopping block when hard times hit. Without an approach that addresses a range of policies and institutions, the government ends up undermining and competing with entrepreneurs rather than serving as their promoter.

How do we know that Iran's business environment in the postrevolution era is the result of the oil windfall and not simply the legacy of the statist takeover of the economy during the war years? Beyond the fact that considerable time has passed since the war's end (more than twenty-five years as of writing), experts argue that Iran adopted less of a "centrally-planned command economy" and more of a "managed war economy" during this period.[190] Furthermore, "Government interventions . . . [during the war] were largely myopic—dictated by immediate force of circumstances and used as short-term crisis management ploys—rather than being based on a well-thought-out and coherent economic plan."[191] We also know that once the war ended, the government set out an "ambitious" reconstruction program that "played a major part in dismantling the country's centrally-administered economy in the late 1980s."[192] In sum, there is little evidence that the Iran–Iraq war permanently shifted the government's approach to the private sector.

How do we know that the nature of Iran's business environment is not, at least in part, explained by the lengthy international sanctions regime imposed on the country during the postrevolution era? We can be confident that this is not the case because the sanctions regime should have led to a less, not more, restrictive business environment over time and especially for the actors who are the focus of this book. Given that sanctions had a negative impact on investors, exporting firms, and domestic firms who were dependent on imported inputs, it would have been logical for the Iranian government to counter this by trying to boost the activities of local entrepreneurs and firms that serve the domestic market. This would strengthen what Ayatollah Khamenei called the "resistance economy" (*eqtesad-e moqavamati*), and improving the business environment for nonelites would have been one way to achieve this. Such logic is apparent in a speech given by Russian leader Vladimir Putin in 2015, after the imposition of sanctions a year earlier. Speaking at a conference organized by the Russian Union of Industrialists and Entrepreneurs, Putin said "greater freedom for businesses is the best response to all external challenges and limitations."[193] If sanctions truly altered how the Iranian government approached the business environment for domestic entrepreneurs and nonelites, we should have seen very different policy making.

4.7 CONCLUSION

The preceding evidence reveals that after Iran received its oil windfall in 1954 the government soon placed numerous restrictions on the activities of domestic entrepreneurs and SMEs. This contrasts with what we observed prior to the oil windfall, particularly in the early Qajar period and during the rule of Reza Shah. Under the oil-rich government of Mohammad Reza Shah, we witnessed a government devoted to utilizing the oil wealth to expand industry, yet the approach was one of ferrying rents and exclusive benefits to a very small group of elites who were handpicked to play a role in this process. Credit allocation and the granting of industrial licenses were two areas in which the government sought to directly control who entered the private sector and who did not. Other economic actors, particularly bazaaris, were prevented from opportunities and certain policies toward them can only be described as antagonistic.

I argued that the cause of this policy approach was not ideological motivations or a way for the government to repress "traditional" and potentially oppositional elements within Iranian society. Instead, it was caused by the oil windfall. The introduction of massive oil revenues into the government's coffers generated intense rent-seeking behavior from elites. For its part, the government provided these rents, protected their business interests, and helped to grow their fortunes, ensuring that they could correspondingly rely on them as a base of political support. Protecting elites, screening market entrants, and erecting a restrictive business environment for domestic entrepreneurs and SMEs was a way to guarantee that those who succeeded in the private sector would be those who helped to bolster the strength of the government.

During the 1970s and into the oil boom, Iran's oil sector rapidly expanded and nonoil industries were steadily crowded out on economic and political terms. The Chamber of Commerce was forced to merge with a Chamber of Industries and Mines, and the resulting organization was unable to influence policy in any real way. The bazaar's guilds were weakened and came under the control of the government. Eventually, the government intervened in the bazaar's activities, sending members to jail and engaging in other forms of repression. It comes as little surprise to find out that entrepreneurs and owners of small firms in the bazaar played an important part in the social movement that removed Mohammad Reza Shah from power.

After the 1979 revolution, Iran faced many new challenges: a brutal war with Iraq, a purging of the political and economic remnants of the monarchy, and a lengthy international sanctions regime. To be sure, all of these affected Iran's economy in ways that can be difficult to quantify. And yet, with respect to the business environment during the postrevolution era, there was a great deal of similarity with what existed prior to 1979. Although the former private-sector elites who prospered during the rule of Mohammad Reza Shah were now gone, we saw the rapid rise of new actors—the *bonyads* and others—to take their place as rent seekers. These organizations have formed a broad base of support for the government, actively defend the government's decisions, and receive considerable autonomy and freedom to capture markets.

Supporting *bonyads* is coupled with a restrictive business environment for private entrepreneurs and SMEs that has led to, among other outcomes, slower rates of robust forms of entrepreneurship in key sectors. The privatization program that is interpreted by many as critical to Iran's economic future has, by most accounts to date, primarily benefited *bonyads* or government-sponsored organizations and not actual private-sector actors. A move to depoliticize the bazaar has led to more monitoring of this group, and conflict between them and the government has erupted on a few occasions. This undoubtedly has worsened their relationship with the state and reduced their chances to affect the contours of policy making. Business associations in postrevolutionary Iran are characterized as weak and/or lacking autonomy, and there can be an oppositional tone to their relationship with the government. They have little capacity to force the government to change policy, much less fully implement the policies it has adopted toward industry.

All of this happened even at a time when we least expected it to. We should have seen Iran implementing business environment reforms during the 1990s. The global oil bust led to a huge drop in oil revenues and the government logically should have looked to the private sector as a potential partner in economic development, given that high oil prices now appeared to be a thing of the past. Furthermore, the oil bust coincided with Iran's rebuilding after the war with Iraq, a time during which the rhetoric increased about looking to market forces in order to help spur development. Here again, the government should have looked to the private sector for help. Finally, the 1990s saw two successive presidents who were known to be strong supporters of reform. Rafsanjani, elected in 1989, was seen as the most pro-private-sector president Iran has had to date. Khatami, elected in 1997, was seen as the most pro-reform president Iran has had to date. The fact that an oil bust, a desperate need to rebuild the economy, and sixteen years of pro-reform governments could not pull

Iran toward reforming the business environment confirms the power of mechanisms set in motion after the oil windfall. Although Iran has faced different challenges and has had different governments throughout the last sixty years, there is an enduring effect of the oil windfall on government policy making toward the private sector.

CHAPTER 5

Doing Business in India's Mining Belt

India is the world's third-largest coal producer and its history of commercial coal production stretches back 240 years. India also has a rapidly growing economy with a vital need to extract the country's energy resources in order to help fuel development. In spite of what seems like an ideal scenario to ask questions about the effects of natural resources, there has been relatively little research on how coal production affects political economy outcomes in India.[1] Beyond what appears in the mainstream press, we know relatively little about the coal sector's impact on government policy making toward the private sector, or how India's coal production impacts the growth of nonmining activities. This chapter is among the first inquiries on the subject for one of the fastest-growing economies in the world.

Beyond the inherent value of analyzing a country like India, there are a number of advantages to using it for this book. An examination of a nonoil resource windfall, such as coal, will tell us about the generalizability of the theory.[2] Indeed, we have little reason to assume that other types of resource windfalls will have a different effect on policy making than oil windfalls. If the resources provoke rent seeking from business elites and policy makers have an interest in controlling the allocation of rents, we should observe a similar effect in India's coal-producing states as what occurred in Iran. In India, mining activities contribute to a federal state's GDP and to state government budgets in the form of royalties paid by mining companies. I analyze India's main coal-producing state, Jharkhand, which is the country's most mineral-dependent state. Mining has accounted for about 15 percent of Jharkhand's gross state domestic product (GSDP).[3]

Another advantage of constructing a case study of India is the fact that its political environment is different from what we see in the majority of oil producers in the developing world. India is a democracy and has been one for a long time. Given the standard argument that many of the negative outcomes in natural resource producers are explained by the presence of an authoritarian government, India allows for a test of the theory in a robustly democratic context. Thus, if we observe a coal windfall having a similar effect in India as it did in oil-rich Iran—two countries with very different political regimes and political histories—we have good reason to believe that regime type is not a critical factor in shaping policy making toward the private sector after a resource windfall. Although the statistical results in Chapter 2 helped establish that the association between oil and a restrictive business environment holds irrespective of regime type, a case-study design has an advantage in being able to more carefully test for the causal mechanisms.

India is different from Iran in another important way: it has a federal structure where states have considerable policy-making powers. During the early 1990s, the Indian government devolved policy autonomy to state governments and granted them greater decision-making power over their policies toward economic development.[4] For more than two decades, resource-rich states in India have carved out different strategies for exploiting their resource sectors and achieving broader developmental goals.[5] Consequently, an analysis of India allows for a test of the theory at the subnational level of policy making. If we witness a coal windfall affecting a subnational government in the same way that an oil windfall affects a central government (as in the case of Iran), we have more reason to accept the validity of the theory.[6]

Finally, and perhaps most importantly for the purpose of theory testing, the unique history of state formation in India allows for a controlled comparison of the effects of a resource windfall as well as the loss of a windfall.[7] Prior to November 2000, Jharkhand was a region located inside Bihar—what I refer to in this chapter as "Undivided Bihar." The Indian Parliament redrew the federal boundaries of Undivided Bihar in 2000, bifurcating the state into (a smaller) Bihar and the new state of Jharkhand (refer to the map in Figure 5.1).[8] With the stroke of a pen, 96 percent of Undivided Bihar's coal and mineral resources were transferred into a newly independent Jharkhand overnight. Due to the fact that the Indian Constitution makes it relatively easy compared to other federal nations to redraw internal boundaries, this provides a truly rare opportunity to test a theory about how a resource windfall—and the loss of a windfall—affects policy making toward the private sector. A comparison of Bihar

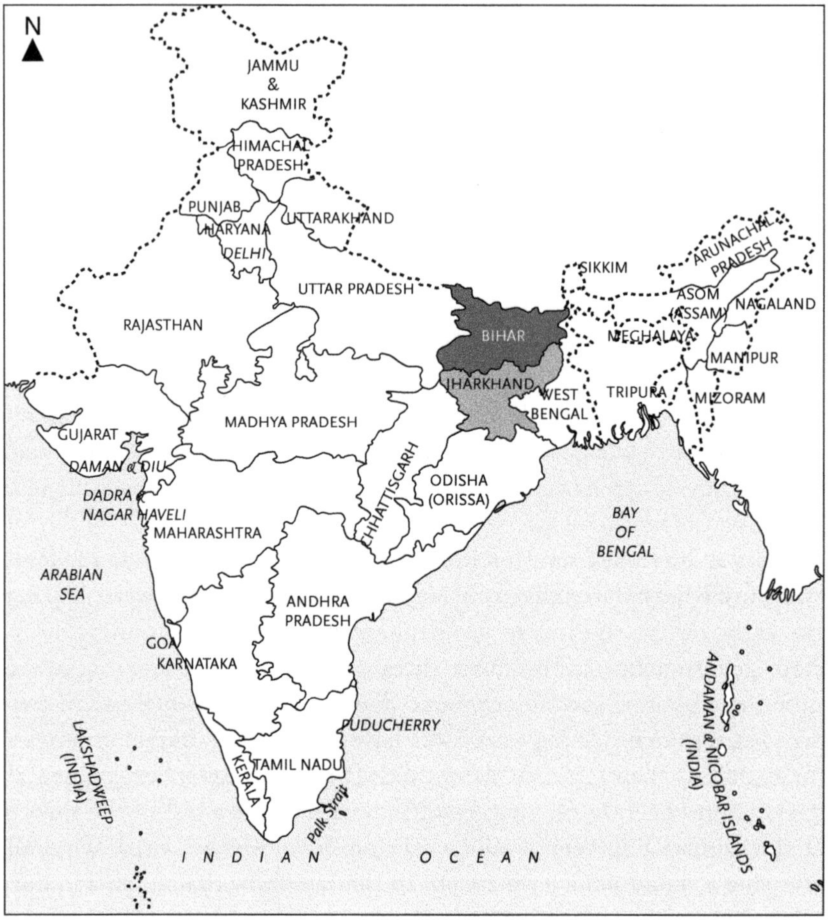

Figure 5.1 Map of India's States and Territories (as of 2012) Highlighting the Locations of Bihar and Jharkhand

and Jharkhand following the bifurcation of Undivided Bihar should tell us a great deal about what natural resources do to a government.

India researchers are only just now beginning to examine what effect this fundamental alteration in the structure of the Indian federation has had on the political economies of these states. Since 2000 to date, India has bifurcated four states to create four new ones, and there are many reasons to be inherently interested in these eight states.[9] Not only can their experiences generate insights about federalism and the political economy of India, but they together comprise 528 million people or roughly 43 percent of the country's total population. The comparison of Bihar and Jharkhand that follows is, to my knowledge, one of the first comprehensive studies about the effects of bifurcation on these two states.[10]

Given the unique histories of Bihar and Jharkhand, there is initial concern that the bifurcation was driven by factors that influence my main variable of interest: a resource windfall. Perhaps India's powerful mining industry lobbied the Central Government to grant Jharkhand federal independence in order to intentionally create a mineral-rich state. Although this theory is plausible, there is no evidence to support it. India experts maintain that the bifurcation was the result of an indigenous (*adivasi*, or "tribe") social and political movement led by the Jharkhandi people for independence.[11] There is considerable evidence of calls for Jharkhand's independence as far back as the 1920s, emphasizing the unique tribal identity of the region. In many ways, the creation of independent Jharkhand was the fulfillment of a dream for a "tribal homeland."

My comparison of Bihar and Jharkhand demonstrates that a coal windfall has a similar effect on policy making toward the private sector as an oil windfall. Jharkhand's business environment since the bifurcation imposes a range of restrictions on entrepreneurs and small firms. The relationship between Jharkhand's policy makers and business elites is based on the exchange of rents for political and material benefits, resulting in the formation of interests that are opposed to reforms. There is little cooperation or sense of unity among small firms in nonmining sectors, and business associations are unable to shape policy making toward the nonmining sector. This contrasts with Bihar's experience since the bifurcation. After losing its coal and mineral wealth, Bihar has ventured down a different policy-making path. Bihar has fostered a business environment that is more conducive to small-scale entrepreneurship, and there is some evidence of a positive impact on private-sector development in the state.

5.1 BIFURCATION AND THE POLITICAL ECONOMIES OF BIHAR AND JHARKHAND

In the decades prior to the bifurcation, Undivided Bihar was classified among India's BIMAROU states, an acronym that translates as "sick" in Hindi. These states possessed the highest levels of poverty, illiteracy, crime, and other socioeconomic ills. Although Undivided Bihar was more industrialized than other states at the time of India's independence in 1947 (largely owed to its mining sector), Undivided Bihar's nonmining industries gradually declined from the 1950s onward.[12] The Central Government provided fewer funds for Undivided Bihar's development compared to other Indian states like Gujarat, Maharashtra, and Tamil

Nadu.[13] With declining nonmining industries and comparatively little in the way of development funds, Undivided Bihar's government became highly dependent on the royalties and cess it received from coal and mineral production. Mining revenues "held the trump card" in the state's finances,[14] and as early as 1981 researchers called attention to a "resource curse scenario" unfolding in the state. There was an "oft-quoted observation that Bihar presents a picture of paradox of poverty amidst plenty."[15]

The calls for Jharkhand's independence from Undivided Bihar were also growing louder during the 1980s. A political movement with its roots in the 1920s was increasingly pressuring the Central Government to grant Jharkhand federal independence. The movement "revolved around the idea that the culturally autonomous indigenous people, or adivasis, were exploited and oppressed by the high-caste Hindu governments that had ruled them from Patna, the capital of Bihar."[16] Additionally, Jharkhandis felt that most of the benefits from their mineral wealth were leaving the area and going north to the Bihar region. Many Jharkhandis reported feeling that they were an "internal colony of Bihar."[17] The importance of Jharkhand's mining sector led it to serve as a site of protest during the independence movement. A blockade of the transport of minerals out of the region in the late 1980s and early 1990s raised public awareness about a crisis unfolding. As the possibility of bifurcation loomed, Lalu Prasad Yadav, Undivided Bihar's chief minister at the time, voiced his opposition to the idea. He famously said that an independent Jharkhand would form "over his dead body."

But on November 15, 2000, the people of Jharkhand were celebrating in the streets. After independence, there was great hope and optimism that the state's mineral wealth would now be used to benefit the local people.[18] For many Jharkhandis, "it was like a dream finally coming true, something that we all had hoped for."[19] Days before the official bifurcation, Madhu Singh, Member of the Legislative Assembly (MLA) from Panki, Jharkhand, boldly claimed: "No one can stop its [Jharkhand's] development now."[20] Anoop Chaterjee, MLA from Nirsa, Jharkhand, predicted: "Even if the Government concentrated 50 per cent on its efforts, Jharkhand would become one of the most developed States of the country."[21]

In Bihar, by contrast, there was fear about the state's economic future given that it lost virtually all of its mineral resources. Bihar state officials complained that the decision to bifurcate was done too hastily and would lead to bad outcomes for average Biharis.[22] Former chief minister of Undivided Bihar, Jagannath Mishra, said the creation of Jharkhand was "the outcome of foolishness."[23] Many pointed to the loss of mineral resources as the primary drawback of the bifurcation.[24] It was reported that some Biharis migrated to Jharkhand in hopes of taking advantage of the economic boom assumed to be forthcoming, and some Bihar state officials requested to be transferred to Jharkhand.[25]

As anticipated, Jharkhand focused its postindependence energy on boosting the state's mining sector. The Jharkhand government's royalties from mineral production are currently the highest in India, as indicated in Figure 5.2.[26] In 2011–2012, Jharkhand's royalties climbed 12 percent higher from the previous year to 1,430.54 crore Rs, or about 270 million US dollars.[27] Coal assumes 90 percent of Jharkhand's mining output and the remaining 10 percent is from iron ore, bauxite, mica, copper, and other industrial minerals.[28] The majority of coal mining in Jharkhand is conducted by the government-owned firm Coal India Limited (CIL). CIL is the world's largest coal company (it is the world's sixth-largest mining company overall) and producer of 81 percent of India's coal. CIL and its subsidiaries, among which are Bharat Coking Coal Limited and Central Coalfields Limited (CCL), operate in Jharkhand. Private mining occurs by Tata Iron and Steel Company (TISCO, today known as Tata Steel) and other producers.[29]

Although the act of bifurcation gave Jharkhand a massive mining sector and associated industries, it left Bihar an industrial shell of its former self.

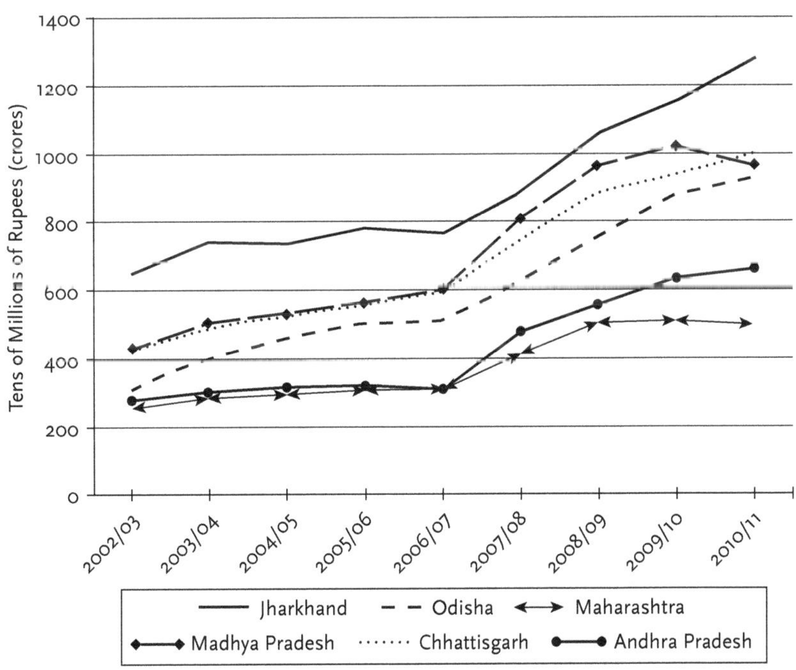

Figure 5.2 Royalties Paid to Selected States for Mining of Coal in India

The majority of Undivided Bihar's large-size factories were all in the mineral-rich regions that spawned the industrial cities of Jamshedpur, Dhanbad, and Bokaro Steel City—all of which were in Jharkhand. Losing Jharkhand was a "huge setback" to Bihar's industry. Prior to the bifurcation, the industrial sector represented 24 percent of Undivided Bihar's GSDP. After the bifurcation, the industrial sector represented just 4 percent of Bihar's GSDP.[30] The legacy of Jharkhand's industrial advantage over Bihar is very much observed today. Jharkhand's manufacturing-based micro-, small-, and medium-sized enterprises (MSMEs) surpass Bihar's with respect to average size, investment, and output. The per unit original value of machinery/equipment for a manufacturing MSME in Bihar is 1.3 lakh Rs, but in Jharkhand it is six times as much at 7.9 lakh Rs. Output value from a manufacturing MSME in Bihar averages 14.6 lakh Rs per unit, but in Jharkhand it is more than twice as much at 34.6 lakh Rs.[31] Bifurcation left Jharkhand with a much more developed industrial sector than Bihar.

5.2 AFTER THE BIFURCATION: POLICY MAKING TOWARD THE PRIVATE SECTOR IN BIHAR AND JHARKHAND

Although Jharkhand received a natural resource windfall because of the bifurcation, since 2000 it has been Bihar who has fostered a more supportive business environment for entrepreneurs and small-sized firms. This is especially the case with respect to government regulation and the adoption of policies directed toward the nonmining private sector. Bihar also has provided more credit to small-sized firms than Jharkhand. The differences observed between Bihar and Jharkhand since 2000 are directly in line with the theory's prediction that the state with the resource windfall (Jharkhand) is more likely to impose a restrictive business environment than the state without the resource windfall (Bihar).

This is evident with respect to the regulatory barriers to starting a business. Data from the 2009 study *Doing Business in India* indicate that Jharkhand's capital city of Ranchi is ranked very low in terms of the ease of starting a business: fifteenth out of seventeen states and territories in India.[32] However, Bihar's capital city of Patna is ranked high at the second spot overall. Based on this one metric, it seems that Jharkhand has a more restrictive business environment than Bihar. But also consider the differences between the states on the measures that comprise this ranking. One area where Patna outperforms all other places in India is with respect to the estimated cost of starting a business. Whereas Patna has the lowest cost in the country, Ranchi is on the higher end and is ranked thirteenth out of seventeen states and territories. Although both states require comparatively more regulatory procedures in order to open a firm, Ranchi requires more than Patna (twelve vs. eleven).[33]

But the actual number of procedures to open a firm in Jharkhand can be higher. A Ranchi-based attorney stated that there are more clearances needed than technically required, and there is significant variation across sectors. There are so many different steps involved that "it would be impossible to make a checklist of them for the starting of any business."[34] Rahul Singh, the Head of the Jharkhand State Office for the Confederation of Indian Industry (CII), said: "For starting a business, there are so many regulations and so much detail involved and paperwork. . . . It is a major challenge."[35]

Yet in Bihar, an accountant saw few difficulties for entrepreneurs in acquiring the various approvals needed for start-up, such as tax registration and industry department approval.[36] Dilip Kumar, a Patna-based Executive at CII, said in 2014 that noticeable changes from the government five or six years ago sped up the process of completing the requirements to open a firm.[37] A Patna-based attorney reiterated that the current business climate reduced previous barriers for entrepreneurs.[38] Arun Agarwal, President of the Bihar Industries Association (BIA), saw the regulatory issues that entrepreneurs face in Bihar getting better "day by day."[39]

Both Ranchi and Patna are ranked near the bottom of the country in terms of the regulatory time delay to start-up, with Ranchi imposing a longer total delay. In Ranchi, the total days of regulatory delay are 104 days, whereas it is seventy eight days in Patna. Rahul Singh observed that in Jharkhand, "the time [delay] is really a bigger problem than the cost of opening a business."[40] For many firms, there are additional delays due to industry-specific issues that need to be addressed, such as infrastructure and access to land, power, and water. Delays can emerge in Jharkhand because it is sometimes unclear which sector the firm is classified in and what specific clearances it needs.[41] A Ranchi-based attorney estimated that it would take six months to get a land permit for a small manufacturing unit.[42] The owner of a manufacturing firm in Jharkhand said it took him two years to get the necessary permits before start-up.[43] A consultant estimated that it would be possible to set up a firm in three to six months, but only if the raw materials are local and little infrastructure is needed.[44] In Patna, however, the general feeling was that time delays were reduced in many areas like industrial permits, and more noticeably if the firm was in a priority sector.[45] Amresh Singh, an owner of a food-processing firm in Bihar, said in 2014 that the time delays have slightly improved but still it can take three to six months of "running around" in order to get clearance.[46]

Analysis of the data in *Doing Business in India* reveals differences in how Bihar and Jharkhand regulate micro- or small-sized firms versus medium- or large-sized firms. When the firm is assumed to be medium, large, or incumbent—and also seeking to expand operations, export, or close down—Jharkhand's business environment imposes slightly fewer regulatory barriers than Bihar's. It is somewhat easier for incumbent, large-sized firms to register property, pay taxes, trade across borders, or close a business in Jharkhand than in Bihar.[47] This is consistent with the theory, given that larger and incumbent firms are more likely to have political connections and are favored by policy makers in Jharkhand. In Bihar, there is no "large-over-small" bias. The state that lost all of its resource wealth needs to support the growth of all types and sizes of firms.

Bihar and Jharkhand differ in terms of credit lending to small-sized firms. Given the statistical findings in Chapter 2 and what we saw in Iran, we expect resource-rich Jharkhand to provide less credit than Bihar. We also expect credit lending in Jharkhand to be biased toward larger and elite-owned firms, similar to Iran after the oil windfall. The data confirm these expectations. Since the bifurcation, banking and financial institutions in Bihar have expanded credit to small entrepreneurs and small-scale industries (SSIs)[48] at a faster rate than Jharkhand. As Figure 5.3 displays, state-owned banks in the two states provided nearly identical amounts of loans and advances to SSIs in the initial years after the bifurcation, but by 2008 state-owned banks in Bihar expanded their lending to SSIs relative to Jharkhand.[49] Loans and advances in Bihar increased by more than 500 percent from 2007 to 2008, while the increase in Jharkhand was much

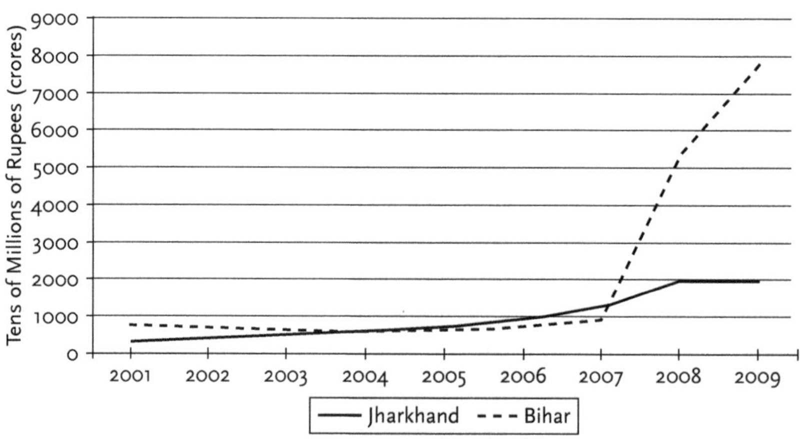

Figure 5.3 Loans and Advances from Public-Sector Banks to Small-Scale Industries in Jharkhand and Bihar

smaller at 58 percent. Lending in 2009 remained the same in Jharkhand, but in Bihar it increased 46 percent.

Other credit-lending data are consistent with a divergence in the two states after 2007. This is witnessed in data on lending from Regional Rural Banks (RRBs), which the Central Government established in 1976 with the goal of boosting the rural economy by developing agriculture, trade, commerce, and industry. RRBs primarily focus on small and marginal farmers, agricultural laborers, artisans, and small entrepreneurs.[50] Initially, most RRBs were established in rural areas but over the years they have expanded to urban areas. Analyzing RRBs is useful because the government controls lending decisions, the ultimate purpose of loans, and a range of other considerations. RRB managers, firm owners, and even the sponsoring banks have "virtually no authority to make any strategic decision."[51] Given this nature of RRB lending, a comparison of Bihar and Jharkhand can tell us a great deal about how each state government approached small-scale entrepreneurs after the bifurcation.

Consider the plots in Figure 5.4 of bank credit from rural RRBs, represented as a share of each state's rural population.[52] During the first five years after the bifurcation, per capita lending in rural areas was marginally higher in Jharkhand. But after 2007, a divergence occurs and Bihar's per capita lending takes off at a steeper trajectory. By 2009, Jharkhand's trajectory mirrors Bihar's, but lending in Bihar remains about 43 percent higher. A more distinct divergence in lending occurs in RRBs located in urban areas, as displayed in Figure 5.5.[53] Bihar entered the 2000s at

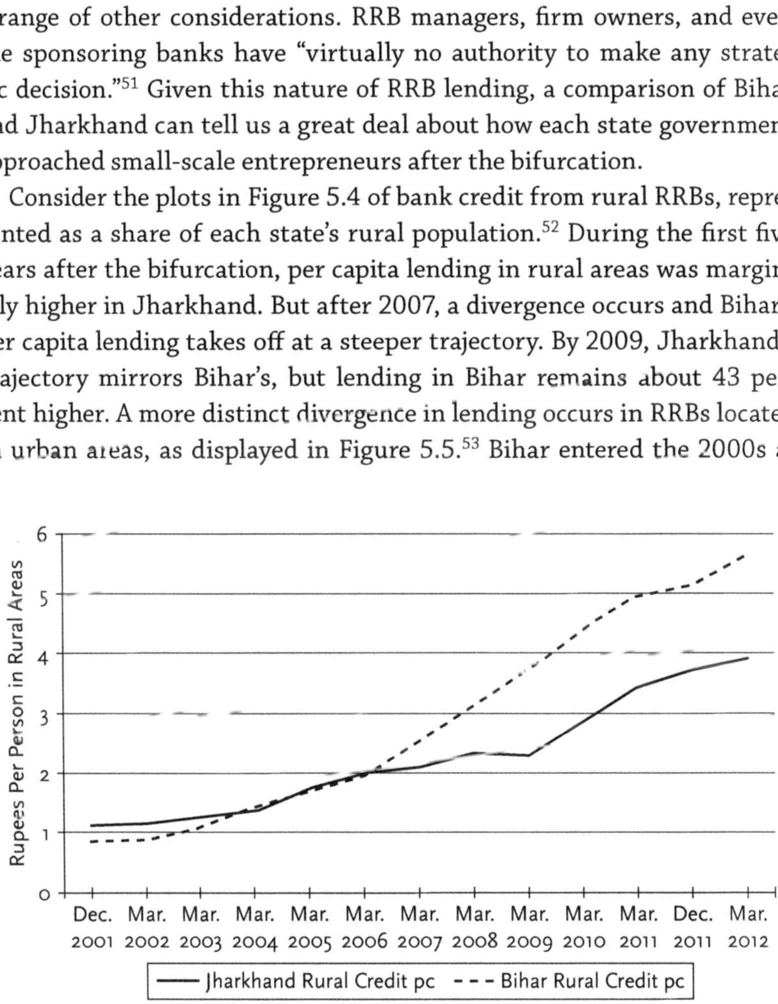

Figure 5.4 Bank Credit from Regional Rural Banks to Rural Areas of Jharkhand and Bihar (as share of rural population)

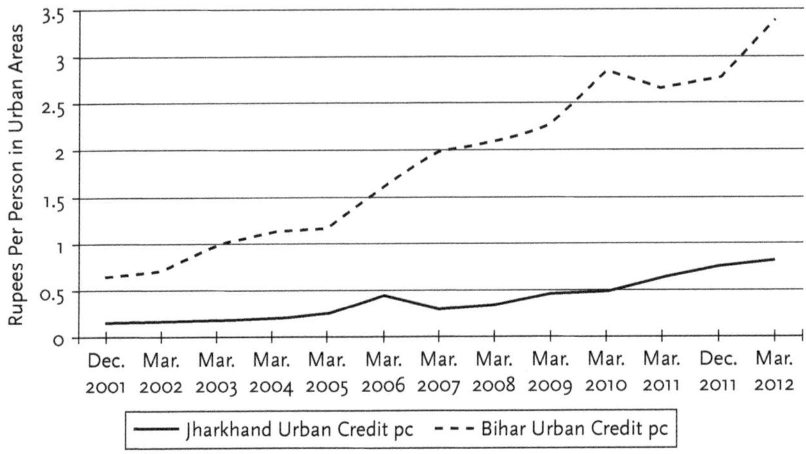

Figure 5.5 Bank Credit from Regional Rural Banks to Urban/Metro Areas of Jharkhand and Bihar (as share of urban population)

a higher level of per capita lending in urban areas than Jharkhand, and since then Bihar has maintained its lead. Per capita lending in urban areas was more than four times higher in Bihar than Jharkhand.

Bihar has made more effort to expand the availability of urban RRB institutions. Throughout the 2000s, as Figure 5.6 shows, the number of rural RRB offices as a share of the rural population declined in both states at nearly the same rate.[54] Yet, the number of urban RRB offices as a share of the urban population increased in both—but much more so in Bihar. This development is all the more notable given that, as of 2011, just 11 percent of Bihar's total population resides in urban areas. Although Bihar's per capita lending to both rural- and urban-based entrepreneurs has expanded since the bifurcation, there is an increased focus on urban areas. In Jharkhand, a higher percentage of the population resides in urban areas, 24 percent, but the level of urban-based RRB lending and number of RRB offices in urban areas lag behind Bihar. Other data reflect a pattern of greater credit availability in Bihar than Jharkhand. It is estimated that 30 percent of Bihar's firms have active bank credit lines, but this value is 13 percent in Jharkhand.[55] Thus, the evidence suggests that Bihar and Jharkhand diverge in terms of credit lending and credit availability since the bifurcation. Jharkhand has underperformed relative to Bihar with respect to providing credit, in particular to small-sized firms and small-scale entrepreneurs. So far, coal-rich Jharkhand looks similar to oil-rich Iran.

Credit access continues to be a major problem for entrepreneurs in Jharkhand. The availability of credit is rated as "extremely low," and this problem is "more severe for Jharkhand's firms than those in most other

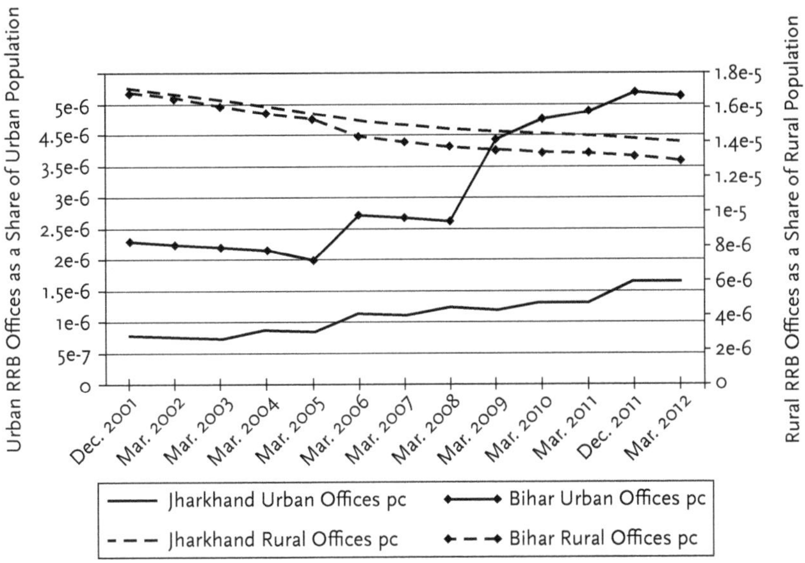

Figure 5.6 RRB Offices in Rural/Urban Areas as Share of Populations

major states."[56] In 2012, Jharkhand's policy makers readily accepted this fact.[57] Many firm owners in Jharkhand view the challenges of accessing credit as more acute for new and small-sized firms. A director of a manufacturing firm in Jharkhand said that it was very difficult for small-sized firms to get credit, but this is what clearly separates the successful firms from the unsuccessful ones.[58] Some said that credit is available to a certain extent in Jharkhand, but this is not because of anything the government has done to promote it.[59] Managers of a business training institute in Ranchi said they can help entrepreneurs apply for credit from state-run banks or specialized microfinance banks, but it is very difficult for entrepreneurs to strike out on their own.[60] A Ranchi-based economist said that it is difficult to get credit unless you are connected to an organization or specific government program.[61] These credit-related challenges were echoed in an interview with Arup Kumar, head of the Jharkhand chapter of the Small Industrial Development Bank of India. Despite the bank's stated mission of providing credit to micro-, small-, and medium-sized enterprises, Kumar said they prefer to loan to entrepreneurs who have "handled such ventures in the past" and were investing more than 10 lakh Rs.[62] But only a very small percentage of entrepreneurs in Jharkhand could make such an investment, given that it is the equivalent of about twenty-nine times the state's per capita income.

Interviews revealed a better credit situation in Bihar, although challenges still emerge and an entrepreneur's experience can vary according to the sector in which they are located. Arun Agarwal of BIA observed that the credit situation has improved overall: "Earlier, banks used to see a huge risk in providing credit to businessmen in Bihar. Now, there is not much of a credit problem."[63] A Patna-based attorney found that getting loans is "much better now than before."[64] Some reported that the credit supply in Bihar is not meeting the demand. A Patna-based accountant said that the government is trying to expand the private sector by providing incentives to entrepreneurs, but banks are not keeping pace.[65] Satyajit Kumar Singh, chairman of the PHD Chamber of Commerce, said that his association constantly pushes the government to increase credit as a way to stimulate local entrepreneurship.[66]

Beyond regulation and credit lending, Bihar and Jharkhand vary in terms of the design and implementation of industrial policy. In Jharkhand, the government is often criticized for adopting policy that does not address the needs of local entrepreneurs. Policy makers are faulted for placing too much focus on expanding the mining sector, with a "lack of vision" for broad-based private-sector development.[67] Many highlight the government's inability to follow through on policies and lengthy delays in the implementation of policy and institutional changes that would ultimately benefit local entrepreneurs. Shortly after independence, the Jharkhand government published its first Industrial Policy document, JIP 2001. Arun Kumar Khemka, President of the Jharkhand Small Industries Association (JSIA), commented that JIP 2001 was "overly ambitious," seeking to grow "every industry in the state and all at once," with "vaguely-worded language and no real direction or mission."[68] The World Bank had a similar reaction to JIP 2001, calling it "unrealistic" and focusing too much on "larger projects, with facilitation of SMEs remaining on the back burner."[69]

In JIP 2001, Jharkhand's policy makers made a bold proposal. They expressed a plan to establish a single-window system (SWS) to streamline business start-up, and would create a new governmental body to manage the SWS called the Jharkhand Industrial Infrastructure Development Corporation (JIIDCO). Yet, sixteen months after the release of JIP 2001 it was dubbed a "dead document." Only 10 percent of the government's stated plans were implemented, JIIDCO was not set up, and entrepreneurs were growing "despondent."[70] Time passed and in 2005, the government again promised to set up JIIDCO and establish a SWS.[71] While JIIDCO finally arrived in 2007, a functioning SWS did not. Five years later, in the long-delayed second Industrial Policy document of Jharkhand, JIP 2012, the government again renewed its promise to establish a SWS.[72] Yet, in December 2012 a headline in *The Times of India* read: "Single Window System a Distant Dream for Jharkhand Inc."[73] As of April 2013, a functioning SWS was not yet in place.[74] Ranjeet Tibrewal, President of the Federation of Jharkhand Chamber of Commerce and Industries (FJCCI), reacted to the continual delay in setting up a SWS:

> We have not seen anything positive in the state in the last 12 years. The single [window] system, which would have helped in transforming the state is still a dream when it is functioning successfully in a number of states including Gujarat *and Bihar*. . . . We feel that no one [in government] is very serious. Entrepreneurs of the state are suffering from problems in getting their files cleared from various departments but there is no one to listen to our misery.[75]

Others criticized the government for its lack of interest in supporting local entrepreneurs. Rahul Singh from CII said, "The government in Jharkhand is not interested in supporting entrepreneurs in the same way that Gujarat or Maharashtra is. But you need this role of the government, through policy, for the business sector to grow."[76] A director of a manufacturing firm said that Jharkhand's government does not support young and nascent entrepreneurs, which is common in other states: "You don't see matching capital . . . you don't see anything towards helping entrepreneurs like you do in other states."[77] The lack of support for entrepreneurs was highlighted by Sanjay Kumar of the Centre for the Study of Developing Societies: "'The [Jharkhand] government has not undertaken any serious initiative to support local industries, especially small-scale industries, and this is kind of surprising."[78] Rahul Singh emphasized that support for small-sized firms is needed to help reduce unemployment, instead of the government's continual focus on large industries.[79] Arun Kumar Khemka of JSIA stated that "The government is the main problem when we are talking about the lack of opportunities [for small-sized firms]."[80]

Bihar's industrial policy, by contrast, is viewed more positively by most business associations and local firms. Unlike Jharkhand, the Bihar government has released a number of official industrial policy documents since the bifurcation—in 2003, 2006, 2011, and a set of revisions took place in 2014–2015. Arun Agarwal of BIA said, "The government is taking interest in the SME sector; they have formulated a good industrial policy."[81] Ashim Mitra, owner of a manufacturing firm in Patna, said that industrial-sector growth is still slow but the government has the "right policies" and is making efforts in the right direction.[82] The 2011 industrial policy was interpreted as the state's most "pro-SME" strategy to date. Principal secretary of industries in Bihar, C. K. Mishra, said that the government was inspired by other fast-growing states in India and set out to formulate a "better policy to attract investment in all category, especially by small and medium enterprises."[83] Dilip Kumar of CII agreed that the 2011 policy "encourages entrepreneurs" and improves the environment for investors.[84]

As Ranjeet Tibrewal noted previously, Bihar has had more success in setting up a SWS than Jharkhand. The government established a SWS in 2006, which imposed a maximum of forty-five days for firms to get clearance for start-up. While this act was important, Dilip Kumar of CII noted that Bihar's SWS currently functions mainly for priority sectors and Satyajit Kumar Singh felt that the SWS was not yet fully implemented.[85] Bihar state officials acknowledge some difficulties with implementation of the SWS and problems in coordination among governmental bodies. In 2014, Naveen Verma, principal secretary of industries in Bihar, pointed to a streamlining of the SWS as a "top priority of the state government."[86] In December 2015, Jai Kumar Singh, minister of industries in Bihar, stated that the government aims to reduce the waiting time to one week for the clearance of all regulations for opening a manufacturing firm.[87]

Since the bifurcation, Jharkhand has imposed more regulatory burdens on entrepreneurs and provided less credit to small-sized firms than Bihar. Jharkhand is often criticized for ignoring the needs of SMEs and focusing instead on large industrial projects. Jharkhand made much slower progress relative to Bihar in implementing key reforms that would help boost entrepreneurship, such as a SWS. Understanding *why* Jharkhand has gone down this policy-making path is the subject to which I now turn.

5.3 POLICY MAKING AND SUPPLY-SIDE FORCES AFTER A RESOURCE WINDFALL: AN ANALYSIS OF JHARKHAND

Jharkhand after the bifurcation is a prime example of how the supply-side forces outlined in Chapter 3—that is, windfall-induced rent seeking, rent-seizing policy makers, and a government that is averse to adopting business environment reforms—unfold after a resource windfall. In order to understand how supply-side forces emerged in Jharkhand, it is necessary to focus on relations between the state government and the primary group of business elites in the state: owners of firms that engage in mining or mining-related activities. In India, private mining firms play an important role in the resource sector through leased government contracts. Although India's coal industry was nationalized in the early 1970s, private firms mine coal through government leases and the production goes to government-owned firms or government-sanctioned users. Private firms also conduct an array of activities that are dependent on the mining sector, such as steel and cement production. Thus, it is important to point out that the sectors in which business elites are most active in Jharkhand include the mining sector. Whereas business elites in Iran mainly focus on nonoil sectors,[88] business elites in Jharkhand can participate in both mining and nonmining sectors.

As the primary site of India's coal reserves and a rich history of commercial mining, private investors should be very attracted to the state of Jharkhand. In the early 1900s, the establishment of TISCO encouraged large industrial investments to come to the Jharkhand region of Undivided Bihar because of its resource wealth. In the 1930s, asbestos and chromite were mined in the district of West Singhbhum, attracting Indian industrial giants such as the Birla family. Coal mining increased throughout the 1980s as companies began operating in the Damodar valley and as iron ore mining grew in West Singhbhum.[89] After the bifurcation, the Jharkhand government negotiated a flurry of memorandums of understanding with large industrial firms, the total value of which stands at more than one trillion Rs, or about 18.3 billion US dollars.[90] A Ranchi-based attorney observed, "Big industries are coming here for the minerals. No one comes to Jharkhand apart from those interested in natural resources."[91] The number of mining projects cleared by the Jharkhand government substantially increased after the bifurcation. In 2008, the area of mining leased was the highest of any state in India.[92]

Immediately following the bifurcation, Jharkhand's government portrayed itself as an active facilitator of investment in the mining sector. In the few years from 1999–2000 to 2002–2003, the number of leases granted increased by 67 percent and the area under mining leases increased by nearly 150 percent.[93] The government helped to broker land deals and established a reputation of not hesitating to alter or ignore mining laws if it benefits investors.[94] Compared to other mining locations around the world, Jharkhand offers fairly competitive tax rates for mining investors.[95] However, preferential treatment toward the sector often involves running roughshod over the rights of tribal communities in mineral-rich areas, displacing villages, violating laws, and ignoring decisions from village-level political bodies.

The Jharkhand government's manipulation of the mining lease allocation process in the coal allocation scam known as "Coalgate" is a powerful example of how rent seizing and rent seeking fosters mutually beneficial relations between policy makers and business elites. Coalgate is perhaps the largest industrial scandal in India's history. In 2012, it was discovered that a select group of mining firms were granted leases through a "murky" and uncompetitive allocation process.[96] While new details continue to emerge as of writing, the scandal took place during the period from 2004 to 2009 and concerned fifty-seven coal blocks, twenty-seven of which were in Jharkhand.[97] An investigation concluded that the mining lease allocation process was corrupt: "A government screening committee chose the recipients; several former and current bureaucrats and industry officials say its decisions were highly subjective, often favoring applicants with close ties to state and national political bosses."[98] Select firms—such as Tata-Sasol, Electro Steel Castings, and Jindal Steel and Power—received coal blocks for free or a small annual fee. Many of these firms then sold the blocks to power and utility companies, or kept the blocks but did not develop them.[99] The Indian Comptroller and Auditor General (CAG) estimated that the outcome of the scandal was that private firms received windfall gains of approximately 1,067,303 crore Rs, or about 200 billion US dollars.

CAG stated that the process of allocating coal mine leases had "opaque" criteria about which firms were granted leases. Although a transparent process of competitive bidding was possible, the Ministry of Coal used a screening committee to make decisions about which firms received coal blocks. State governments provided opinions on companies that applied for blocks, they would advocate on behalf of certain firms, and officials received various rewards for their efforts.[100] The case of Abhijeet Group is one example. In 2005, Abhijeet Group received four coal blocks in Jharkhand and in 2008 they received an additional block. During both instances, P. P. Sharma was Jharkhand's chief secretary (the highest-ranking government officer in the state), who advocated on behalf of Abhijeet Group. After Sharma's retirement in 2010, he was appointed a director at Abhijeet Group. While such a practice is not uncommon, the Sharma case "suggests that senior bureaucrats may have thrown their weight behind companies that rewarded them post-retirement," and points to the "links between politicians and companies."[101] Reflecting on the Coalgate scandal, justices of the Indian Supreme Court said that officials misrepresented firms to the screening committee,[102] and "a large number of beneficiaries were either politicians or their relatives or associates."[103]

Preferred firms received a staggering number of benefits during the Coalgate years. The Jayaswal business family—out of which the Abhijeet Group and Jayaswal Neco Industries were formed—was given mining leases for a total of ten blocks (six of which went to Abhijeet Group, and five of these six were in Jharkhand). This totaled more than nine hundred million tons of coal. And yet, the blocks were awarded at a time when the group had just one coal project in operation and Jayaswal Neco Industries had been issued a court injunction due to unsatisfactory progress on its mining projects.[104] A reasonable question asked afterward was how one family business could secure so many coal blocks when it had little experience in the sector and little progress made on existing projects. The answer "could lie in the family's political clout. . . . Abhijeet Group seemed to have the ability to win the support of bureaucrats."[105]

Another inquiry by India's Central Bureau of Investigation (CBI) revealed that the Jindal Group "misrepresented facts" and was "shown favour by the Jharkhand government" for the allocation of coal blocks it received in 2007.[106] Although the Ministry of Power was initially against giving the blocks to the Jindal Group, a former minister of state for coal, Dasari Narayan Rao, helped sway the ministry in favor of the Jindal Group. The CBI stated that, in exchange for his efforts, one of Rao's companies received an investment bribe of 2.25 crore Rs, or about 350,000 US dollars.[107] While ongoing investigations continue to uncover the extent of the Coalgate scandal, experts say that this type of high-level corruption in the mining lease allocation process was well known before news of the scandal broke.[108] One said that it was an inevitable feature of allowing the private sector to play a role in mining, a problem that the 1973 nationalization act was designed to solve in the first place.[109]

Madhu Koda, Jharkhand's chief minister from 2006 to 2008, was implicated in the scandal and was imprisoned from 2009 to 2013 under charges that include illegally granting mining leases.[110] While he was in office, the CBI uncovered a scam whereby a private-sector lobbyist, Nira Radia, bribed Koda to illegally extend a mining lease for Tata that was about to expire.[111] Koda also recommended allocation of thirteen coal block leases in Jharkhand, but a CBI investigation revealed "irregularities" in this process.[112] Three of the coal blocks were granted to dummy companies formed just prior to the allocation.[113] Koda helped facilitate the granting of mining blocks to Rungta Mines, Sunflag Iron and Steel Ltd., and Electro Steel Castings, all companies implicated in the CAG's Coalgate report.[114]

Coalgate and the activities of Madhu Koda are not the only examples of Jharkhand's policy makers providing selective rents to business elites. In an unrelated scandal, it was reported that the government granted TISCO a 999-year lease, in violation of Section 8 of the 1957 Mines and Mineral Development and Regulation Act (MMDR), whereby the maximum period of a mining lease is not to exceed thirty years. This gave TISCO a lease to mine with a rent of Rs 3,250 per year, which is the rent they paid in 1946 and is the equivalent of about 60 US dollars.[115] Another instance concerns the links between Madhu Koda and Tata Steel in the allocation of an iron ore block, which violated the MMDR and the 1960 Mineral Concession Rules Act. The Indian government's Revision Authority of the Ministry of Mines referred to "large scale irregularities during Madhu Koda regime and illegally allotted iron ore mines of Tata Steel . . . all the decisions taken by Koda Government . . . should be scrutinised and action be taken against guilty fellows."[116]

When asked why Coalgate occurred, EAS Sharma, secretary in the Ministry of Power from 1997 to 1998, said: "Politicians realized that this kind of coal was black gold."[117] Indeed, Jharkhand's policy makers manipulated the mining lease allocation process and provided rents to business elites because they reap political, professional, and material benefits from doing so. Some were given direct cash payments for their efforts, others were provided lucrative positions on the boards of large companies, while others enjoyed the ancillary benefits that accompany having close ties to India's captains of industry. A Jharkhand-based economist observed, "The situation is out of control. Rent-seeking politicians, other rent-seekers from industry, there are multiple actors who are trying to gain rents in Ranchi."[118]

Political donations are another motivating factor to provide rents to business elites. In India, politicians rely on donations to win elections and private mining or mining-related firms are the main contributors to the country's major political parties. From 2004–2005 to 2010–2011, political parties in India received more than 4,500 crore Rs, or about 820 million US dollars, from donations and contributions.[119] A study by the Association for Democratic Reforms (ADR) found that twenty-six of the top fifty companies who made donations to the Congress Party are linked to mining or other resource extraction activities. Companies implicated in the Coalgate scandal—such as Jindal Group, Adani Group, Shyam Group, Essar Group, and others—were highlighted in the ADR report, as were companies that provide logistics or services to the mining sector.

For instance, Jindal Group gave 1.05 crore Rs, or about 192,000 US dollars, to the Congress Party. Essar Group gave 75 lakh Rs, or about 137,000 US dollars, to the Congress Party. Other firms that are directly or indirectly dependent on the mining sector, such as Torrent Power and Vedanta, generously gave to the Congress Party's main rival and the current party in power, the BJP.[120] What is especially noteworthy about the ADR report is that Indian companies located in sectors whose activities *do not* require state governments to provide permission or licenses contributed almost nothing to political parties. In fact, "The biggest donors were involved in mining, power and other sectors dependent on the government to obtain rights to natural resources."[121]

As Jharkhand's policy makers maintain a system of exchanging rents for benefits, this has predictably resulted in a development strategy that emphasizes the mining sector—the sector that benefits business elites the most. An alternative strategy that emphasizes broad-based private-sector development through business environment reforms is therefore unlikely to come about in Jharkhand. Given the pattern of relations between policy makers and business elites in the state, there are many reasons why a policy maker would not want to risk being identified as opposed to the mining sector and supportive of broad-based private-sector development—irrespective of whether or not such reforms would impact business elites in a negative way. In a highly competitive democracy such as India's, a policy maker in a mineral-rich state that sends a message of opposition to the mining sector may inflict irreparable damage on their own career, not to mention the electoral prospects of their party. Such a policy maker is unlikely to last long in a context like Jharkhand's today. The resource windfall has fueled rent seeking and entrenched relations between the government and business elites who are dependent on the growth of the mining sector. This leads policy makers to adopt interests that are opposed to reforms that signal a more balanced approach to private-sector development.

Jharkhand's frequent rotation of chief ministers in the postbifurcation era reinforces the urgency with which policy makers need to establish their support for business elites linked to the mining sector. In the nearly fifteen years since its independence, Jharkhand has had six different chief ministers and twelve changes in power.[122] Prior to the term of the chief minister as of writing, Raghubar Das (assumed office on December 28, 2014), the average length of time in office for any chief minister in Jharkhand was quite short: 503 days. Although the cast of individuals who dominate Jharkhand's politics are familiar and two major parties contest for control—the BJP and the Jharkhand Mukti Morcha (JMM)[123]—state policy makers have a small window in which to cement their status as decision makers and/or reap benefits from holding office.

This sense of urgency may be partly due to India's democratic nature. Ross argues that politicians in democracies are more likely to allocate rents at a faster pace than politicians in autocracies because they have a harder time granting exclusion rights (given the division of formal powers) and the practice of competitive elections makes their power less secure.[124] The speediness with which Jharkhand's policy makers entered office and then became embroiled in a coal allocation scam is just one example of this. Shaibal Gupta of the Asian Development Research Institute argues that the highly contested, volatile nature of Jharkhand's politics is one of the main factors that leads to policy-making uncertainty in the state.[125]

Given this essential nature of Jharkhand's politics since the bifurcation, a policy maker's rational approach is to adopt interests that support business elites and push for a continued expansion of mining and mining-related activities. While a shrewd policy maker is one who always pays lip service to supporting new entrepreneurs and firms of all sizes, the presence of a resource windfall and entrenched rent seeking may make it a serious political risk to promote policies that intend to spearhead business environment reforms. Policy reforms that encourage broad-based private-sector development are unlikely in Jharkhand because they will not benefit politicians and mining elites, who have a large influence on policy making.[126] A Jharkhand-based economist explained, "Reform does not happen because the mining sector . . . [has] interests they want to protect."[127]

5.4 NONELITES AND DEMAND-SIDE FORCES IN BIHAR AND JHARKHAND

In Jharkhand, business associations that advocate on behalf of SMEs are generally ineffective at influencing policy makers and cannot realistically compete with the interests of business elites. As the preceding suggested, business association leaders in Jharkhand have a generally unfavorable view of government policy making to date. But they have little capacity to affect changes in policy, as their role is primarily one of criticism using media interviews or the release of studies and documents. R. D. Singh, secretary general of Federation of Jharkhand Chamber of Commerce and Industries, said that he was occasionally asked to provide input on policy planning but felt that his suggestions were not taken seriously.[128] Jharkhand-based leaders of national business associations, like CII, and regional ones, like FJCCI and JSIA, all argued that the government is the main cause of barriers to entrepreneurship and the reason for slow growth in nonmining sectors. They also lamented the frequent rotation in chief ministers as creating an environment that suggests governmental instability and policy-making uncertainty.[129] In Bihar, by contrast, interviews revealed a closer relationship between business associations and policy makers. Satyajit Kumar Singh, chairman of the PHD Chamber of Commerce, said: "We are in continuous talks with the government about regulation."[130] Arun Agarwal stated that his organization, the Bihar Industries Association, is "invited to many decision-making forums. We have fairly good relations with policymakers. They listen to us and accept our advice sometimes."[131] Overall, the capacity of business associations to influence policy makers appears to be weaker in Jharkhand than Bihar.

In order to get a sense of how the business environment affects SME owners and understand the degree to which owners have the capacity to influence policy makers, I conducted interviews in the wood furniture sectors of each state. In Jharkhand, I sampled formal firms from every third storefront encountered in two areas of Ranchi: Kadru Road and Bahu Bazaar, Chutia. In Bihar, I used the same method to approach firms in two areas of Patna: Nala Road and Kankerbagh. The wood furniture sector was selected for a few reasons. One is that the sector is dominated by small firms producing for the domestic market. Most have less than ten employees and their customers live nearby. This allows me to obtain information on firms that are smaller than the ones examined in *Doing Business in India*. Raw materials are available locally or within India, and thus entry into this sector should be easier than others. The wood furniture sector does not compete with the mining sector in Jharkhand or the agricultural sector in Bihar, and thus we would not expect the business environment to be unusually restrictive. Neither state government has a special initiative to boost the wood furniture sector; consequently, the environment is more likely to be representative of other nonprotected sectors.

Owners in Jharkhand were more likely than those in Bihar to talk about the difficulties of start-up. The owner of a firm in Ranchi reflected, "Starting a business is difficult here, but if you have the motivation to do it then you can deal with all the headaches."[132] Another said, "It is very difficult to start a new business here, especially if you need skilled staff."[133] A door maker in Ranchi, who has three employees, agreed: "It is difficult and expensive to start a new shop here. . . . It has gotten more expensive over time."[134] In Bihar, owners more often pointed to how easy it is to open a firm and how this has resulted in a more competitive business environment in recent years.[135] One owner said, "I have lived on Nala Road since 1964. . . . There are now 25 people selling the same thing in the same small area when there was just a few. In the last four years, we have seen a lot of change."[136] Although more competition put a different kind of stress on Bihar's owners, a few said that this was evidence of an economic boom in the state.[137] Unlike in Jharkhand, virtually no owner in Bihar reported difficulties in locating unskilled and skilled labor.

In spite of the increase in competition in Bihar, none said that this had a negative effect on relations between firm owners in the sector. Yet, in Jharkhand, owners talked about the lack of cooperation. One owner in Jharkhand said, "[U]nity among furniture makers would help set prices. That way everybody can get what they deserve rather than fighting against each other. But there is no real sense of unity or community among the furniture makers."[138] The issue of mistrust came up repeatedly in Jharkhand, in spite of the fact that most businesses are family run and many have been in operation for decades in the same location. Another owner in Jharkhand stated, "There are many problems in the business sector here. Unity is an impossible issue to solve because of the competitive nature of the furniture market. It doesn't facilitate friendly relationships between companies."[139] Overall, interviews in Jharkhand suggested that firms in the wood furniture sector are fragmented from one another, and it is difficult to achieve coordination among firms to deal with issues related to the business environment.

When asked about their interactions with the government, an owner in Jharkhand responded: "I have had minimal interactions. . . . My store is registered, but apart from that I have no relationship with the government and I have not received benefits from any government program."[140] The owner of a business who supplies wood to furniture makers recalled receiving a loan from the government five years ago, but similarly stated that he "has no relationship with the government and has never been a part of any initiative."[141] Another owner was recently forced by the government to move his shop without compensation due to a street-widening project. He said, "The government is always bothering me for something or other."[142] Another owner commented that the government's bad policies for forest resources have made it increasingly difficult for him to access raw materials, forcing him to import wood from Myanmar, Nigeria, and European countries.[143] By contrast, owners in Bihar said that acquiring raw materials for their firm was relatively easy.[144]

When asked about the availability of credit in Jharkhand, one owner said that this is an area of "difficulty" and that "reputation is the most important thing for accessing credit."[145] The owners of a firm, which has remained open for three consecutive generations, agreed with this statement. The youngest of the owners said, "We have a good reputation and so it is easy to get credit."[146] However, an owner of a smaller operation (three employees) said it would be difficult for him to get a loan because he rents his space and does not have any collateral. He described the credit application process as "tedious with a lot of paperwork."[147] A door maker recollected that he was unable to get credit when he started.[148]

In Bihar, credit was less a hot-button issue and only 21 percent of those interviewed had ever applied for a loan. As in Jharkhand, enterprises with three workers or less said they either did not need a loan for such a small operation or did not have enough collateral to qualify.[149] One firm in Bihar was in the process of applying for a loan and remarked that it was "difficult because they require so much paperwork and you have to hire a lawyer. . . . The process has been ongoing for the last two to three months."[150] Two other owners in Bihar applied for loans from state banks in 2008. One estimated that it took about a month to process everything and the other owner estimated that it took about six months.[151] The owner who waited six months added that he was "lucky" because he had a "good manager" in charge of the application and did not have to pay a bribe.[152]

The issue of extortion repeatedly came up in Bihar. Most owners reported that they used to pay money to the local mafia for protection but that this practice has now stopped. One owner said, "I had to give to the local mafia a few years ago but after the new government [of Nitish Kumar] came into power that problem went away slowly. Now you will rarely hear about that type of extortion."[153] Most attributed this change to the government's recent efforts to stamp out the mafia.[154] In Jharkhand, firms reported that corruption, bribery, and extortion are simply part of doing business. Rahul Singh of CII said that firms "need to set aside a fund in order to bribe local middlemen and the mafia. Extortion of firms is a big problem in this state. It is a reality for doing business and it is not only in the rural areas but also in Ranchi."[155] Rural areas of Jharkhand are particularly prone to these activities. An analyst said that entrepreneurs need to have a relationship with the mafia, Naxalites, or local bureaucrats in order to open a firm in rural Jharkhand.[156]

Taken as a whole, interviews with business association leaders and SME owners in the wood furniture sectors revealed differences in Bihar and Jharkhand. Whereas business associations in Bihar are actively involved in discussions with policy makers, in Jharkhand they have a more distant relationship with the state and largely function as critics. Although the patterns in each state's wood furniture sector are only suggestive of nonelite firms as a whole, there was little indication that owners in Jharkhand have the capacity to shape the government's policies that constrain their activities. Interactions between firms and the government were reported in a negative light in Jharkhand, but less so in Bihar. The lack of cooperation among owners in Jharkhand was understood as a barrier to creating a more profitable and stable environment in which to conduct business. Compared to Jharkhand, owners in Bihar appear to operate with less government involvement in their activities. In Bihar, the government was praised more than it was condemned, particularly with respect to improving law and order, infrastructure, and policies that help boost private-sector development.

5.5 AN ALTERNATE EXPLANATION IN JHARKHAND: POLITICAL INSTABILITY AND THE COAL MAFIA

An alternate explanation for Jharkhand's more restrictive business environment is the role of political instability. Indeed, most India experts would rank Jharkhand as having an unstable political situation, both in terms of formal politics and local state-society relations. Jharkhand, along with other states in the eastern region (Andhra Pradesh, Bihar, Chhattisgarh, Odisha, and West Bengal), has faced an ongoing struggle with a Maoist insurgency, referred to as the "Naxalite movement." Jharkhand also faces instability caused by the activities of a coal mafia and an illegal coal trade, and to some extent this has overlapped with the Maoist insurgency in recent years.[157] The Jharkhand government's inability—or unwillingness, as many argue—to control the illegal coal trade and the coal mafia reinforces political instability at the local level and may dampen policy makers' interests in pursuing an alternative development approach that would deemphasize the mining sector.

The precise volume of illegal mining in Jharkhand is unclear but reported to be enormous. Some estimate that illegal coal mining affects as much as seventy to eighty million tons of coal annually.[158] The monetary loss to the state each year is estimated to be about 600 crore Rs, or about 110 million US dollars.[159] This amount is roughly half of what the state received in total coal royalties in recent years.[160] Illegally mined coal is transported in Jharkhand primarily by truck and bicycle. The coal is stolen from existing mines, produced from illegal mines, or taken from trains after they are loaded. There is evidence of illegal mining at opencast mines where industrial projects have since moved on, and even on land directly adjacent to government-run mines.[161] Violence in the illegal coal trade is commonplace as wars erupt between rival mafia groups.

Many argue that the illegal coal trade is facilitated by bribery and extortion in the form of "kickbacks to the coal managements and the local administration to avoid penal action."[162] Others hold that it occurs through the assistance of railway police and officials who help transporters give coal to the mafia for storage and future sale.[163] Yashwant Sinha, India's finance minister from 1990 to 1991 and 1998 to 2002, told the Indian Parliament about how the mafia extorts money by overcharging buyers when they load purchased coal: "Lakhs of rupees exchange hands everyday.... Coal India, CCL officials are directly involved. I discovered that every evening the spoils are divided in the GM [General Manager]'s office. The local police, too, are involved in this."[164]

To be sure, Jharkhand's coal mafia was not born after the bifurcation. Jharkhand simply inherited the coal mafia because it inherited the coal industry. The coal mafia engaged in similar activities in Undivided Bihar, where "close links" existed between the coal mafia, politicians, and the bureaucracy.[165] Reflecting on the problem today, a Jharkhand-based economist said that the illegal coal trade continues because "it benefits everyone. Mining companies support illegal mining ... they see this as a small price to pay for their mining. The local politicians support illegal mining because they receive a portion of the revenues as extortion."[166]

Here again, we observe a relationship based on exchanging rents for benefits between governmental actors and a group of business elites: in this case, the coal mafia. On the surface, it appears that material benefits are provided to mostly lower-level governmental actors for helping to facilitate—or turn a blind eye to—the illegal coal trade. But there is evidence that the illegal coal trade has involved higher levels of state government, both during the period of Undivided Bihar and after the bifurcation. Shibu Soren, three-time chief minister of Jharkhand and former coal minister of India, once tried to remove a coal ministry officer who reported on corruption and theft at CIL's most productive mine in Chhattisgarh.[167] A Jharkhand researcher stated that the coal mafia has a "very important role in Jharkhand's politics and especially in local elections. They will help elect certain people to government, and everyone knows they are supported by the coal mafia, and that is just how it is."[168]

The continuation of the illegal coal trade requires that the coal mafia has access to Jharkhand's resources, whether to steal or illegally mine. In this sense, the coal mafia is similar to the business elites discussed in the preceding text in that it has a clear interest in seeing Jharkhand pursue a development strategy that furthers the expansion of the mining sector and not one that deemphasizes the role of mining to achieve broad-based growth. A link exists between the coal windfall, political instability at the local level, and the formation of local interests that favor a focus on the expansion of mining activities. This may help to reinforce the dynamics described previously that lead Jharkhand's policy makers to continually focus on the mining sector.

5.6 CONCLUSION

Since the bifurcation, the resource windfall in Jharkhand has negatively affected the state's business environment for local entrepreneurs and SMEs. The presence of a mutually beneficial relationship between policy makers and business elites has resulted in interests that are opposed to reforms that would encourage broad-based private-sector development. Insights from business association leaders revealed that formal avenues of influencing policy makers are generally ineffective, and information from firm owners in the wood furniture sector suggested that cooperation among SMEs is likely to be low. The presence of local-level instability and the illegal coal trade may further discourage a policy strategy that promotes broad-based private-sector development. Bihar, by contrast, has fostered a more supportive business environment for local entrepreneurs and SMEs since losing its resource windfall in the bifurcation. Bihar has imposed fewer regulatory barriers on small-sized firms, provided more credit, and implemented more industrial reforms that benefit local entrepreneurs. Bihar's business associations engage more with policy makers, and firm owners in the wood furniture sector appear less frustrated with the government's policies and the current business environment.

Bihar has seen more SME growth than Jharkhand since the bifurcation, which may suggest that its policy approach is already generating results. Consider the differences in the number of industrial SSIs that were registered in each state before and after the bifurcation.[169] Figure 5.7 displays the ratio of the number of industrial SSIs registered in Bihar relative to Jharkhand.[170] In 1995, the Bihar region of Undivided Bihar registered about 1.2 SSIs for each one registered in the Jharkhand region.[171] Following the bifurcation, Bihar maintained a significantly higher ratio year after year. Bihar registered about 2.7 industrial SSIs for every one registered in Jharkhand. Data on the registration of new privately-owned firms are consistent with this pattern. In 2000–2001, Bihar registered 285 new private firms. This number dipped to 243 firms in 2001–2002, but steadily rose each year to 1,271 firms in 2010–2011. Unfortunately, similar data are unavailable for Jharkhand from 2000–2001 to 2005–2006. However, a comparison of the two states from 2006–2007 to 2010–2011 indicates

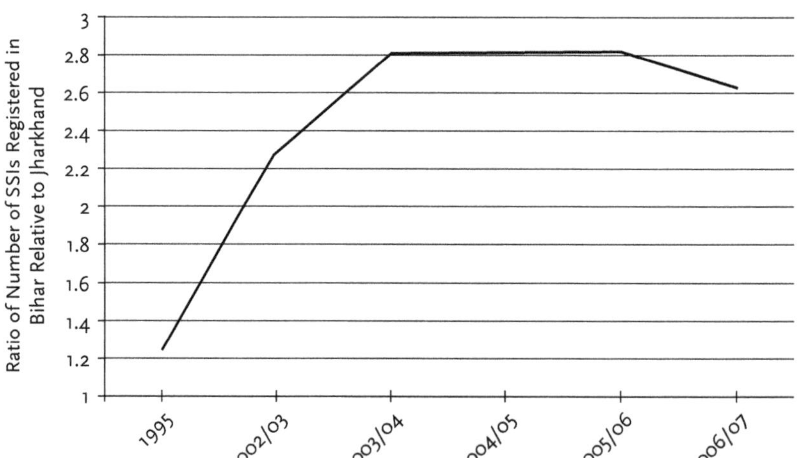

Figure 5.7 Number of Small-Scale Industrial Units Registered in Bihar Relative to Jharkhand (before and after bifurcation in 2000)

that Bihar registered more than twice as many new privately-owned firms than Jharkhand (4,294 vs. 2,128).[172]

A comparison of income per capita growth reveals that Jharkhand's industrial advantage over Bihar on the eve of the bifurcation did not translate into a sustained developmental advantage afterward. This is seen by examining the best fit loess curves in Figure 5.8 (the dots in the figure represent the raw data).[173] During the first four years after the bifurcation, Jharkhand (solid dots) averaged higher average growth relative to Bihar (hollow dots). But after 2006–2007, Jharkhand levels off and Bihar starts to accelerate. From 2006–2007 to 2013–2014, Bihar averaged higher annual growth compared to Jharkhand (9.6 percent vs. 7.3 percent, respectively) with less volatility overall (the standard deviation is 4.4 vs. 7.7, respectively). Although more time surely needs to pass before we can make any claims about their overall developmental trajectories, Jharkhand's relatively weaker performance after 2006–2007 begs the question: In achieving independence did Jharkhand also inherit an economic resource curse? In 2007, the World Bank saw "some signs of [Jharkhand] developing the 'resource curse' syndrome."[174]

Much as it is too soon to declare Jharkhand "resource cursed," it is misleading to speak of Bihar as a state firmly on the path to economic transformation. Although Bihar has made important strides in the policy areas discussed in the preceding text, and additionally with respect to law and order and infrastructure,[175] many contend that Bihar had nowhere to go

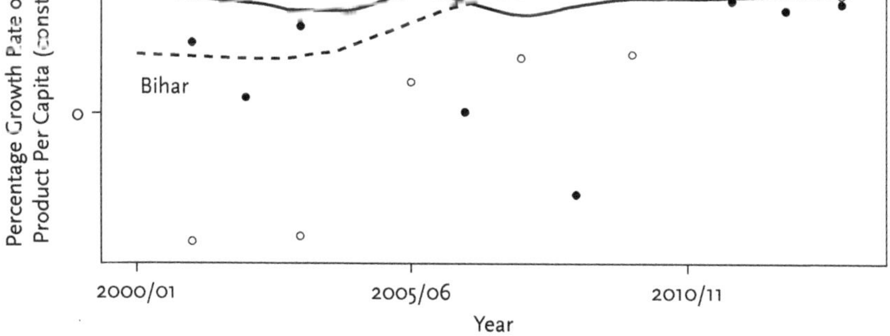

Figure 5.8 Bihar and Jharkhand, Loess of Annual Growth in Income Per Capita (at constant 2004–2005 prices)

but up.[176] For that matter, some argue that Jharkhand simply inherited all of the corruption and mismanagement in Undivided Bihar and there is no reason to believe it would not respond positively given the right leader.[177] Undoubtedly, it will be important to continue watching how the political economies of Bihar and Jharkhand unfold over time. Comparing the two can tell us a great deal about what happens when one state receives a huge resource windfall while another state loses one.

CHAPTER 6

Reform in the Gulf Arab Region

Saudi Arabia since the 1990s

In Chapter 3, I laid out a probabilistic theory about the effect of oil windfalls on the interests of policy makers and elites about business environment reforms and the capacity of nonelites to demand reforms. The analysis of Iran revealed how the microfoundations of the theory arise in a long-term oil producer over time. The analysis of Jharkhand demonstrated that the theory applies when the natural resource in question is coal, the country is a full democracy, and policy makers are operating at the subnational level. The stories I told about Iran and Jharkhand paint a picture of resource producers falling into a "trap" of a restrictive business environment that renders them vulnerable to a host of negative economic, political, and social outcomes. Stagnating nonoil incomes, slow rates of entrepreneurship, bribery of public officials, and industrial sectors relying on outdated technologies—these are just some of the outcomes that result after receiving a resource windfall.

But some places might avoid this trap. As I discussed in Chapter 2, the Gulf Arab oil producers emerge as weak outliers to a general trend of higher oil income and a restrictive business environment. Teasing out the mechanisms that explain the ability of a select group of oil producers to adopt business environment reforms—even limited and tentative ones—is the goal of this chapter.[1] In order to do so, I focus on the case of Saudi Arabia. Throughout the last fifteen years, Saudi Arabia has initiated reforms that foster a more supportive business environment than what we see in the vast majority of oil producers in the developing world. In 2013, the World

Bank declared Saudi Arabia the MENA region's highest-ranking economy on the overall ease of doing business, placing it globally at the twenty-second spot among 185 countries.[2] While rankings are only suggestive and some argue that Saudi Arabia's record of reform is less remarkable, there are many signs that Saudi Arabia has made progress toward improving the business environment for domestic entrepreneurs and SMEs.

Saudi Arabia is not the only Gulf Arab oil producer to adopt these reforms, but it is the most logical case to select for this study. In terms of geographic size and population, Saudi Arabia looks similar to other developing countries while the other Gulf Arab oil producers are all geographically small, sparsely populated states. In terms of oil production, Saudi Arabia eclipses all of them by volume, but it is second-lowest among Gulf Arab oil producers in terms of per capita oil income in recent years. Saudi Arabia is closer to the median oil producer on this measure than a place like Qatar, for instance. Finally, the case of Saudi Arabia has intrinsic value. It is undeniably the country most associated with the word *oil*.

In order to explain Saudi Arabia's ability to initiate business environment reforms, I center on the country's monarchical system[3] and the institution of economic familism—two factors that are broadly present in the other Gulf Arab oil producers who chart similar pathways of reform. I argue that Gulf monarchies are able to pursue business environment reforms because the presence of a stable dynastic monarchy compels them to continually legitimize and reinforce the familial ties between policy makers and the private sector as well as within the private sector. The Saudi monarchy's reinforcement of the institution of economic familism provides a guarantee to business elites that their rents and interests will continue to be protected over the long run. Consequently, business environment reforms are less likely to directly threaten elites' interests and, instead, may bring them greater economic gain.

As I argued in Chapter 3, one of the principal reasons why elites are opposed to reforms is because there is *uncertainty* about how reforms will affect their rents. Given the nature of these types of reforms, business elites are concerned that lowering the barriers to competition and providing opportunities to new entrepreneurs may strip away or devalue the rents they strive so dearly to hold on to. The case of *bonyads* in Iran are illustrative of this point. *Bonyads* have received such a wealth of benefits and privileges from the Iranian government that they are opposed to reforms that would limit their economic activities or open the economy to outsiders. Protecting their rents is of paramount concern and business environment reforms may threaten their rents. What makes Saudi Arabia different from Iran is that it has a stable monarchical system that continually reinforces the institution of economic familism. This serves as a mechanism that reduces uncertainty about the effects of reforms on the core interests of business elites. In Saudi Arabia, reform is possible because there is a mechanism that continually protects business elites' rents even in the face of reform.

6.1 MONARCHY, THE COMMITMENT PROBLEM, AND THE ROLE OF ECONOMIC FAMILISM

In order for policy makers to get business elites on board with reforms, they need to convince elites that they are committed to protecting their interests throughout the reform process. At the minimum, policy makers need to convince elites that they will respect the terms of existing contracts and the rights of private property holders. Ideally, they can convince them that their rents and privileges will be fully protected and their interests will not be negatively affected by opening up the market to new entrepreneurs and investors. This show of commitment is perceived as more credible if there is a binding mechanism that can force the government to honor elites' interests into the future. To be sure, the challenge in an authoritarian context, such as Saudi Arabia's, is that even with such a mechanism in place the country's policy makers can ignore or overrule it whenever they want. In this scenario, business elites have little reason to trust policy makers and are more likely to try and block reforms in order to preserve the status quo.

I argue that Saudi Arabia's monarchical system can solve certain types of commitment problems that are needed to initiate economic policy reforms. Although I am the first to make this argument to my knowledge, it directly builds from insights we already have in the literature. Studies on the Middle East demonstrate that monarchies persist through "consensus" among the royal family and a group of elites unrelated to the family but bound to the regime by the patronage they receive.[4] We also know that a system of hereditary succession can benefit regime-tied elites because it maintains their relative power in the country.[5] The role of regime-tied elites is crucial to understanding the inherent advantage that monarchies have in instituting policy reforms. Lisa Anderson argues that in monarchies "the availability of support from the old elite may allow substantially greater levels of *voluntary compliance* with the state's new extractive ambitions than would rule by 'the people' or 'the proletariat' that alienates the old elite before a new one is available to take its place."[6] The fact that a study of twenty-seven OECD countries found that during the nineteenth and twentieth centuries monarchies were more likely than republics to implement sweeping reforms—rather than incremental ones—may be explained by the greater policy-making flexibility inherent in monarchies.[7]

But there needs to be some basis for the "voluntary compliance" and "consensus" that exists between a monarchy and elites. Indeed, if the monarchy exercises absolute rule, as in Saudi Arabia, why would elites simply trust that it will continue to protect their interests during policy reforms? I argue that monarchies rely on the institution of *economic familism* to solve the commitment problem between policy makers and elites, which helps to facilitate the adoption of business environment reforms. In the presence of absolute rule, rent-seizing policy makers, rent-seeking elites, and weakly enforced or absent formal institutions, an institution of economic familism protects the interests of elites over the long run. In Saudi Arabia, the monarchy reinforces the belief that familial bonds in the economy—whether directly to the royal family or through family-owned businesses—provide a guarantee to elites.[8]

Economic familism is a system whereby familial ties underpin economic activities and directly shape the norms of interaction between economic actors. The importance of familism in economic interactions and institutions is observed in many places around the world—from the Appalachian United States to China—though the form that economic familism takes in each place varies according to historical experience and cultural interpretations of the concept of family.[9] Common attributes of economic familism include the following: kinship relations are the primary way of financing and facilitating business transactions; family-owned businesses are given preference over non-family-owned businesses; nepotism is a preferred practice in hiring, promoting, and transferring assets; intermarriages commonly occur between established business families; and preferences about financial partnerships are determined by kinship relations. Where a system of economic familism dominates, familial ties guide entrepreneurs in everything from how to raise capital to how to maintain clients. In some contexts, familial ties essentially predetermine a firm's fate in the economy.

Economic familism is not the same as "crony capitalism." Crony capitalism concerns "insider" relations between individuals and the government (which may or may not involve familial ties or nepotism) and is interpreted in a negative light as one specific form of corruption. Although a system of economic familism can foster crony capitalism as one outcome, in a fundamental sense economic familism is a culturally defined institution with ambiguous outcomes that involve a far broader array of interactions beyond the relationship between "insiders" and the government. Economic familism can shape all economic transactions by highlighting biological foundations, which has one common manifestation in the prevalence of family-owned firms. Furthermore, not all places that rely on the institution of economic familism are affected by crony capitalism. The economies of Germany and Sweden, for instance, are highly dominated by family firms but they also score low on corruption indices. These economies are fundamentally different from the classic examples of crony capitalism that were highlighted in the East Asian Crisis, such as Indonesia and Malaysia.[10] Finally, crony capitalism tends to reinforce policy stasis whereas economic familism can coexist with a program of economic reform. As Francis Fukuyama points out, "Those who, like Max Weber, argued that Chinese familism would impede economic modernization were simply wrong."[11]

6.2 THE EMERGENCE OF BUSINESS ENVIRONMENT REFORMS IN SAUDI ARABIA

It is necessary to understand the extent to which Saudi Arabia has embarked on a program of business environment reforms before discussing how the monarchy uses economic familism to solve the commitment problem with business elites. The first traces of reform emerged in Saudi Arabia's development plans after 1985. At this point, the government started to pursue a "more holistic approach to the economy" and embraced a "new vision" for development.[12] Motivated in part by a desire to join the World Trade Organization (it became a full member in 2005), the Saudi government established new institutional bodies that provided the foundations for a less restrictive business environment. Some argue that Saudi Arabia's economic approach since 2000 is "similar in some respects to that of the developmental states of East and South-East Asia,"[13] with the sense that a "new business climate" is transpiring in the country.[14]

New institutional bodies, particularly the Supreme Economic Council (SEC) and the Saudi Arabian General Investment Authority (SAGIA), were created in the months between August 1999 and April 2000. Whereas the SEC would evaluate the effectiveness of economic, industrial, agricultural, and labor policies, SAGIA was tasked with promoting foreign investment, attending to the business community's interests, and acting as a mediator between the state and private investors.[15] Just prior to the creation of these bodies, surveys revealed that business owners viewed the procedural requirements to obtain a business license as "tedious" and taking up to six months with no guarantee that a license is awarded.[16] Applying for credit was viewed as a time-intensive and challenging process, one that varied between banks and government sources. After the birth of SAGIA (an intended "one-stop shop" for the business community), the regulatory barriers to establishing a business and obtaining licenses were reduced. Although this now made it easy to acquire a business license (at least initially), the start-up process was still far from streamlined. Other regulatory requirements—for instance, obtaining commercial registration—continued to impose barriers.[17]

While Saudi entrepreneurs still face challenges, the government's initiation of reforms make the country stand out from its oil-producing colleagues around the world. By 2011, the World Bank's *Doing Business* reports stated that Saudi Arabia reduced the regulatory procedures required to obtain construction permits; the regulatory procedures and cost of registering property; and the regulatory procedures, bureaucratic delay, and official fees required to open a business.[18] Five years before, Saudi Arabia ranked near the bottom of all nations in regulatory and financial barriers to opening a business, and far below the global median with respect to protections for investors.[19] Yet, some Saudi Arabia experts criticize the *Doing Business* reports as exaggerating the degree of reform that has taken place. As a comparison, the World Bank's Worldwide Governance Indicators suggest a more modest degree of reform in regulatory quality from 1996 to 2014. Measured on a scale of -2.5 to 2.5, Saudi Arabia's lowest score over this period was -0.2 in 1998 (42nd percentile) and its highest score was 0.18 in 2010 (56th percentile).[20] But other measures rank Saudi Arabia among the developing world's top reformers. The World Economic Forum relied on surveys of business executives to generate a "Global Competitiveness Index" that ranks Saudi Arabia eighteenth out of 144 countries in 2013.[21]

Other government-led efforts convey an interest in sparking domestic entrepreneurship and private-sector growth. The government provides many incentives to private start-ups, such as interest-free, medium- and long-term loans; subsidized leasing at industrial sites; preferential treatment for locally manufactured goods in procurement contracts; and a number of other benefits.[22] A member of the Riyadh Chamber of Commerce and Industry stated that the government is trying to increase training opportunities for local entrepreneurs and encourage more research on entrepreneurship.[23] King Saud University's Riyadh Techno Valley is training students in innovation and business planning while offering financial support for top students to implement their business ideas.[24] The creation of a commercial credit bureau that catalogues the credit exposure of Saudi firms was lauded as a step toward increasing information about local corporate credit and debt.[25] The establishment of the Capital Market

Authority (CMA) regulatory agency in June 2003 was interpreted as helping contribute to a "financial revolution" in Saudi Arabia.[26] After initiating its work in 2004, the CMA "got off to a quick start."[27] Steffen Hertog observes, "bankers generally recognize it as a more qualified and committed regulator than most other Middle East stock market authorities."[28]

Much like Iran, the privatization of state assets is a contentious topic in Saudi Arabia. Unlike Iran, however, privatization has proceeded slowly but steadily since the 1990s. In areas where it has taken place, it has generally been successful. The areas in which the private sector is most active include the postal services, ports, telecommunications, insurance, aviation, banking, health, education, water and wastewater management, and mining.[29] Surveys show high levels of support for privatization from the general public, the business sector, and some members of the royal family and senior bureaucrats.[30]

Privatization and the reforms outlined in the preceding text have resulted in Saudi Arabia's private sector playing an increasingly important role in the economy. As Figure 6.1 shows, the private sector's percentage of GDP increased after 1968 while the oil sector's percentage of GDP declined.[31] The increasing role of the private sector from 2001 to 2008 is a particularly strong indicator of the government's commitment to the private sector, given that oil revenues rapidly grew throughout these years.

More new and smaller-sized firms have entered Saudi Arabia's economy since the start of reforms. Although annual data on the number of new SMEs is not available, looking at how the average firm size and assets have changed over time provides insights into the growth of SMEs.[32] Figure 6.2

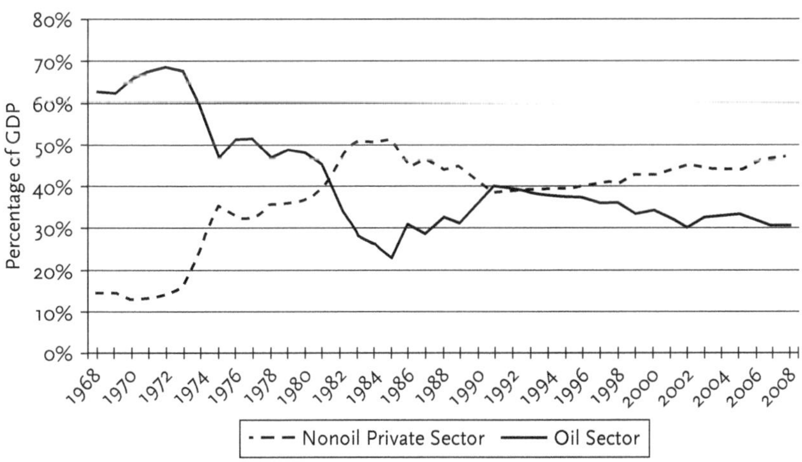

Figure 6.1 Saudi Arabia's GDP, Oil Sector and Nonoil Private Sector

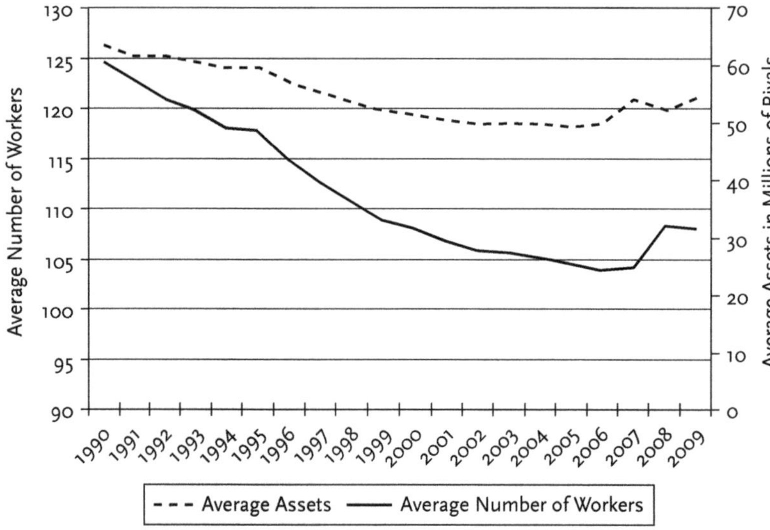

Figure 6.2 Nonoil Industrial Units in Saudi Arabia, Average Number of Workers and Total Assets, 1990–2009

plots the change in the average number of workers and total assets of firms in the country's nonoil industrial sector since 1990.[33] Whereas the average firm in the nonoil sector employed about 125 workers in 1990, this number steadily declined afterward (solid line, left y-axis). By 2009, the average firm had 108 workers. This development contrasts with the 1970 to 1990 period, during which the average number of workers employed per factory grew rapidly.[34] Given that the total number of nonoil industrial units increased from 2,065 to 4,433 over the time period 1990–2009,[35] it is unlikely that other factors (such as enhanced productivity) caused the downward trend in average firm size. Instead, a large number of new, relatively smaller firms started to enter Saudi Arabia's nonoil sector after 1990. The other plot in Figure 6.2 helps to corroborate this finding. The average assets of firms in the nonoil sector decreased over time (dotted line, right y-axis), which suggests that new, smaller firms with relatively less capital entered the market.[36] We also know that from 2005 to 2007 there was a 21 percent increase in the number of municipal business licenses issued and a 24 percent increase in the number of municipal construction permits awarded.[37]

Despite the government's initiation of reforms, Saudi Arabia experts emphasize that considerable work remains to be done. Indeed, the government still makes it easier for foreign- or joint-owned firms to succeed in the market than domestic-owned firms,[38] and yet it continues to place regulatory barriers on the former's activities.[39] Large firms are routinely favored over SMEs in terms of credit lending,[40] as indicated in a 1999 report that found that 75 percent of all bank loans are given to large firms.[41] While the Saudi government makes occasional efforts to increase financial support for SMEs,[42] it is below an appropriate level given the importance of SMEs in the economy.[43] For instance, the value of guarantees that the Saudi Industrial Development Fund provided to SMEs from 2006 to 2009 was a fraction of what it provided in loans to larger industrial units.[44] According to Abeer Allam, this is because "[t]raditionally, banks felt more comfortable lending to big business families whose senior members usually sit on their boards, while small and medium-sized business owners struggled to get finance."[45]

Nevertheless, there is optimism about what the future holds for SMEs in Saudi Arabia. In 2013, the economist John Sfakianakis said, "More credit will be available for SMEs. It has already changed and it will change [further]."[46] In 2010, the HSBC Small Business Confidence Monitor reported that Saudi SMEs have the highest level of business confidence in the Middle East and second highest in the world.[47]

6.3 THE POSSIBILITY OF REFORM: RENT SEEKING AND FAMILY FIRMS AFTER THE OIL WINDFALL

The Saudi government's reforms were made possible by the presence of a stable monarchical system that continually reinforced the institution of economic familism. This served as a mechanism that reduced uncertainty about the effects of business environment reforms on the core interests of business elites. Understanding how a monarchical system and economic familism facilitated reform in Saudi Arabia requires us to go back in time to see the roots of rent seeking and the growth of family-owned firms after the oil windfall.

Saudi Arabia first started to export large amounts of oil after World War II. From this point forward, many individuals in the country began engaging in rent-seeking behavior and thus were able to carve out important roles for themselves in economic and political realms.[48] Hertog describes this as the "dizzying years" of the 1950s and 1960s, when "uncontrolled, Byzantine expansion based on patronage" allowed business elites to expand their operations and friends of the monarchy made forays into the private sector. A number of now-powerful business families emerged this way, starting out as preferred suppliers or serving as advisers to Saudi Arabia's first monarch, Ibn Saud. Some of them were given "exclusive trade agencies" by the monarchy, which they retain even today.[49]

The rise of these elites as a result of oil-fueled rent seeking produced a contingent of powerful family-owned firms in the country. Today, many of these firms are gigantic conglomerates that dominate the sectors in which they operate and hold a privileged position over firms with less storied histories. A few prominent examples are the Alireza, Khashoggi, Juffali, Jomaih, Pharaon, and Rajhi families.[50] These families conduct activities across a wide range of economic sectors, from banking to construction to defense. The Al Jomaih family, for example, began business activities in the 1930s and is now a multi-billion-dollar conglomerate active in automotive services, manufacturing, soft drink bottling, real estate, and more. The Pharaon family made billions in banking, tourism, and petroleum sectors, all the results of the initial activities of Rashad Pharaon, personal physician and adviser to the royal family. By the mid-1980s, Rashad's son Ghaith was reportedly the chairman of the board for twenty-two corporations.

One of the oldest and most famous family-owned business empires is the Alireza family from the Hijaz region. They began operations in 1845 as an importer of foodstuffs, and one branch of the family business became the first company on the peninsula to import and distribute automobiles for commercial purposes. Now a massive conglomerate, the Alireza Company is active in real estate, shipping, luxury imports, and more. Xenel Industries, a company founded in 1973 by members of the Alireza family, is the firm that was responsible for restoring electricity to Kuwait City at the end of the first Gulf War.

After the 1950s, another group of family-owned firms emerged alongside this group of older family-owned firms. Many of these new entrepreneurs came directly from the government bureaucracy and opened businesses that became family-owned conglomerates over time and through the government's support.[51] These more recent arrivals to the private sector allied with the old family-owned firms described in the preceding text, and established the private sector as the place for bureaucratic elites to turn toward for opportunities after leaving government. In many cases, bureaucratic elites turned toward the private sector before leaving government.

An alliance between the older family-owned firms and those owned by bureaucratic elites was solidified through intermarriage and the preferred practice of transferring ownership of assets from father to son.[52] Intermarriage among family-owned firms in Saudi Arabia has reinforced both the familial underpinnings of these firms and the salience of kinship ties throughout the private sector. Intermarriage among family-owned businesses helps to build trust between previously unconnected firms. This has led family-owned businesses to "have the tradition to do business among themselves,"[53] which further strengthens the salience of familial ties in the private sector. Unsurprisingly, studies of Saudi businesses find that managers are primarily loyal to their own family and kinship groups.[54] Yet, intermarriage between members of family-owned firms does not always yield optimal outcomes. An ongoing financial scandal concerns familial ties between the Al Gosaibi family conglomerate and Maan Al Sanea, owner of the Saad Group, who married Sana, sister of Saud al Gosaibi. After their marriage, Al Sanea was accused of defrauding the Al Gosaibi family conglomerate of 9 billion US dollars.[55]

Today, family-owned firms play such a substantial role in the Saudi economy that, in the eyes of many, they wholly define the country's private sector. Up to 90 percent of all companies in Saudi Arabia are family owned.[56] In 2013, their contribution to GDP was estimated to be 25 percent.[57] Family-owned businesses reportedly hold the equivalent of 70 billion US dollars in investments, and the majority of trade agencies and franchises in the country are owned by roughly one hundred family-owned businesses.[58] Most of these firms have been around since the initial days of rent seeking after the oil windfall. In fact, "very few big newcomers" have entered the country's private sector as a whole since the 1980s. Most of the big stakes in the economy (as in the bureaucracy) were already "parceled out" by then.[59] Thus, the rents and business interests of family-owned businesses were already established and supported by the Saudi monarchy well before the onset of business environment reforms.

The Saudi monarchy maintains a hands-off approach toward regulation of family-owned businesses, which reinforces the salience of economic familism as an institution. The government has yet to impose regulations on family-owned firms but instead encourages "internal reform."[60] Many highlight the lack of transparency among family-owned firms, a phenomenon that emerges not just in Saudi Arabia but also throughout the Gulf region. Family-owned firms, as a whole, have yet to embrace practices that are now routine in many countries, such as corporate governance, disclosure, and transparency.[61] As a result, their business practices tend to be described as "quixotic."[62]

The Saudi monarchy reinforces the institution of economic familism by giving family-owned firms preferential treatment with respect to credit lending. Family-owned businesses are generally favored by banks and lending institutions. Intermarriage between powerful family-owned firms boosts each one's creditworthiness. In the case of the Al Gosaibi and Al Sanea families noted previously, intermarriage facilitated cross-family guarantees and was a "key factor" in getting access to credit.[63] An advertisement in the *Washington Post* on Saudi Arabia's liberalized business environment boasts: "Creditworthiness and stability are two things that the second generation can also count on when taking over the business from their fathers, which come in handy if a company needs to raise money for future expansion or diversification projects."[64]

The Saudi monarchy's preferential treatment of family-owned businesses reinforces the role of economic familism as an enduring institution in the economy. This institution sets norms for what types of firms receive priority over others, which in turn directly shapes the decisions of investors and asset holders. The advantage of family-owned businesses over other firms is simply a reality of how business is done in Saudi Arabia. Indeed, "To an outsider, it would be difficult to discern who is related to whom by marriage or blood affiliation, but for those operating in the Kingdom this is more transparent and it influences how business is carried out."[65] As the monarchy has committed itself to supporting family-owned businesses over time, the fear about the effects of policy reforms on these elites' interests correspondingly subsides. The rents of family-owned firms are protected.

6.4 THE POSSIBILITY OF REFORM AND THE ROYAL PRIVATE SECTOR

Economic familism serves as a binding commitment for another key segment of the Saudi private sector: members of the royal family who engage in business activities. Convincing this group of business elites that their interests and rents are protected during a reform process is critical, and there is no reason to assume that the monarchy will automatically satisfy all of their interests. In spite of the monarchy's longevity (more than eighty years), the state has had to deal with many conflicts within the royal family and disagreements about policy matters. Yet, the vast majority of princes and other royal family members are not members of the "strategic elite," and they have no real input on policy making. Many of these individuals entered the private sector after they "realized they would have no chance of advancing through the [political] system."[66] Thus, the monarchy needs to provide a guarantee that these royals' business interests are protected during a program of business environment reforms.

There are numerous examples of royal family members who joined the private sector, particularly during the oil boom. As Saudi Arabia's wealth grew from its oil production, "The lines between state finances and individual wealth became blurred as some princes used political office for personal financial benefit."[67] This blurring of politics and finances also led to an increasing division between royal business elites and the rest of the private sector.[68] Mishaal bin Abdul Aziz, son of Ibn Saud, is one of the more notable members of the royal family who entered business early on and achieved success. He founded the Al-Shoula Holding Group in 1970, a multi-billion-dollar firm whose main activities are in real estate development, with a recent expansion into petrochemical services. It was reported that control of Al-Shoula was handed over to Mishaal's son, Prince Abdulaziz bin Mishaal, during the late 1990s. Other sons and grandchildren of Ibn Saud have achieved success in the private sector. Nawwaf bin Abdul Aziz is a major investor in the real estate and energy sectors. Other royal family members include Mohammed bin Fahd, Faisal bin Salman, and Abdul Aziz bin Fahd.

Undoubtedly, the most famous Saudi royal family member to make a mark in business is Prince Al-Waleed bin Talal. He is the richest Saudi Arabian and, as of 2015, the thirty-fourth richest person in the world. He has amassed a fortune from real estate and investments in Citicorp and News Corporation, among others. Prince Talal's investments have in turn benefited other members of the royal family, such as his sister Sultana Nurul Al-Waleed, who reportedly owns 16 percent of the Savoy Hotel in London, which Talal bought in 2005.

The individuals mentioned in the preceding text are examples of sons, grandsons, or granddaughters of Ibn Saud who entered the private sector. When we widen the scope of inquiry to include nephews and more distant relatives we observe a similar pattern. Some of this expansion of the royal family's business activities was due to the fact that Faisal, half-brother and successor to Ibn Saud's son Saud, extended benefits to members of his own family branch who were technically not related to Ibn Saud. Although the distribution of rents under Faisal was "relatively well controlled . . . his rule was no exception to the basic pattern."[69]

It is important to note that royals are rarely alone in their economic activities but instead are surrounded by "growing numbers of hangers-on, advisers, and business partners."[70] As a result, the influence of royal business elites extends beyond the actual royal family. This network allows nonroyal entrepreneurs to gain a major advantage if they are able to get a royal patron on board with a project. Thus, royal business elites are not the only ones who have benefited from rent seeking and who have built up their business activities over time: a layer of nonroyal entrepreneurs are capitalizing on this branding of the "royal private sector" through personal, familial, and business relationships. Unsurprisingly, the ability to establish a solid link to the Saudi royal family can make a remarkable difference in terms of an entrepreneur's access to rents.

Drawing from within the royal family to create a new segment of the private sector blurs the lines between government and private sector, as noted, but it also creates a group of principled, family-tied elites for whom the commitment problem is easily solved—provided that an institution of economic familism is present. Because of their direct familial ties to the monarchy and the salience of economic familism, these elites are more likely to perceive that their rents and interests will continue to be protected during the reform process and they will logically be the first to benefit from new economic opportunities. Business elites from the royal family often block attempts by other groups to dominate sectors of the economy and prevent outsiders from gaining excessive economic influence.[71] There are many reports of royal business elites abusing their power to gain advantage over nonroyal members of the private sector.[72]

Although the vast majority of royal business elites do not hold decision-making authority, some "business-minded princes" are influential in shaping policy making in recent years.[73] The creation of the SEC in 1999 helped boost the policy-making influence of a select group of business elites from the royal family.[74] Familism also plays an important role in the development of the Saudi bureaucracy. Hertog finds that, "Saudi bureaucracy building often seems not to be a case of 'form follows function' but of 'form follows family' . . . power politics within the family—whatever its details—consistently loomed larger than other considerations when it came to negotiating institutional change."[75] As a select group of royal business elites continue to shape policy making and bureaucracy building in Saudi Arabia, the institution of economic familism grows stronger.

It is difficult to envision an institution other than economic familism that is capable of solving the commitment problem between the monarchy and royal business elites. Given that the longevity of the monarchy is, in part, determined by its ability to coordinate and satisfy the needs of the royal family members who compose it, providing a commitment to royal business elites is synonymous with preserving the monarchy. Implementing business environment reforms that directly threaten the interests of royal business elites could contribute to a destabilization of the monarchy. Implementing reforms that improve the business environment—while still protecting the interests of royal business elites through legitimizing and reinforcing the institution of economic familism—is possible. Without an institution to solve the commitment problem, however, royal family members with substantial investments have reason to fear that business environment reforms will threaten their interests.

6.5 CONCLUSION

The Saudi monarchy's reinforcement of economic familism solves the commitment problem with business elites and thereby facilitates the initiation of business environment reforms. Without the presence of economic familism as an enduring institution, business elites living in a context of absolute rule and entrenched rent seeking driven by the oil windfall are likely to oppose these reforms. They fear that reforms will increase market competition and strip away or devalue the rents they strive to hold on to.

Although Saudi Arabia has made progress toward reforming the business environment, a number of issues remain inadequately addressed. The country has an abysmal record when it comes to supporting female-owned businesses (even compared to its Gulf neighbors), despite some exceptions among the royal family or traditional business families.[76] In addition, most Saudi firms engage in petty corruption and bribery in order to smooth their relationships with officials,[77] though this practice is common in many countries (and especially those with oil wealth, as Chapter 2 illustrated). Furthermore, questions remain about whether reforms will lead to a group of entrepreneurs with an organizational identity. Saudi business, as a class, has become "much more mature" over time and relative to the bureaucracy, but it still remains limited in its ability to shape policy making.[78]

Saudi Arabia's experience with business environment reforms leads us to question the long-standing argument in the social sciences about a "king's dilemma," whereby monarchies are inherently unable to pursue modernization without risking their own downfall.[79] Although the concept of modernization is highly contested and has lost its appeal among scholars in recent years, to the extent that business environment reforms are one facet of *economic* modernization, the Gulf Arab oil producers are exceptions to the king's dilemma. While the king's dilemma may still apply to political or social modernization, a sustained program of economic reform has the capacity to spark reforms that go beyond the market and into society and politics.[80] Entrepreneurs and firm owners may be economic actors on a fundamental level, but they are also social actors who can press political demands on leaders as we saw in Iran during the time of the 1979 revolution. Given a commitment on the part of the government to expand opportunities for the private sector, entrepreneurs may come to shape political and social reforms in the Gulf Arab oil producers over time.

The case of Saudi Arabia leads us to wonder about how alternate institutions might substitute for economic familism and solve the commitment problem in *non*monarchical, authoritarian oil producers. Some suggest that successful autocrats can solve the commitment problem by playing the role of the "stationary bandit," in which they rationally provide public goods while not overextracting from the population.[81] But the stationary bandit solution suffers from practical and theoretical failings: "The practical problem is that no one lives forever. The time horizons of despots are not infinite. . . . The theoretical problem is that the despot's commitment . . . is purely volitional."[82] Other potential solutions include constitutionally limiting the government's power[83] or creating an independent body (such as a guardian judiciary) that prevents policy makers from violating the rights of investors and asset holders.[84] Reputational norms, a hostage mechanism, or third-party enforcement[85] could help guarantee the rights of business elites in an authoritarian context.

Basically, there are many ways, theoretically speaking, that an authoritarian state could solve the commitment problem with business elites and convince them that their rents are protected during a reform process. But we do not know if any of these alternate ways will actually function in a nonmonarchical, authoritarian context. Saudi Arabia may be unique in that its policy makers were invested in making reforms and they had a way to commit to the interests of business elites. Should policy makers in other authoritarian oil states desire to adopt business environment reforms, they will need to convince business elites to get on board with the idea or policy makers need to reduce their political dependence on business elites. The more time that passes, however, the more difficult it will be to achieve either of these outcomes.

CHAPTER 7
Conclusion

In 1991, King Fahd of Saudi Arabia reportedly said, "A Saudi has to be very unlucky, very stupid and very lazy not to do well."[1] One major goal of this book has been to show that an analysis of government policy making tells us more precisely why some individuals in oil-producing countries fail to do well (even if they also happen to be unlucky, stupid, and lazy). Since the start of the world's oil boom, the governments of oil-producing countries have collectively failed to foster business environments that support the growth of entrepreneurs and SMEs. Instead, governments have centered on providing rents to private-sector elites with whom they have direct ties. Even when such a group was not present (as after the 1979 revolution in Iran), a new one was quickly fashioned. Those in the private sector without ties to the government are disadvantaged in two respects. Not only are they forced to compete with rent-seeking elites, but also they encounter a government that adopts policies that impose additional barriers to their activities.

In one sense, the prevalence of rent seeking after an oil windfall may be less of a chronic problem if oil producers would enforce market competitiveness and "live and let live" when it comes to the core activities of doing business. Those who obtain rents will rise faster and find ways to threaten or force others out of the market, but we might view fewer disparities in the private sectors of these countries. We should also view a less pronounced stagnation of nonoil income in these countries over time. A large part of the problem is that oil producers adopt policies that weaken those who are *already weak* by definition. This occurs on economic terms, but it has effects that reach to the realm of politics and society. Many of the negative effects of a restrictive business environment were outlined in Chapter 1, but here I provide discussion about three additional areas of importance: female entrepreneurship, corruption and trust in the government, and opportunities for SME growth and investment.

A lack of entrepreneurial opportunities, high barriers to formal entry, and the politicization of credit lending will disproportionately affect female entrepreneurs. Around the world, female entrepreneurs are more likely than male entrepreneurs to open informal firms instead of formal firms. Yet, informal firms tend to have lower productivity, lower income-earning potential, and fewer (if any) legal protections. Studies show that female entrepreneurs tend to have worse access to credit than male entrepreneurs.[2] Thus, we have reasons to believe that female entrepreneurs in oil-producing countries will face major challenges—firstly, due to widespread biases against women in business and, secondly, due to oil's effect on regulatory and credit-related barriers to entrepreneurial activities. The lack of women-owned firms in the formal sectors of oil-producing countries will reinforce the gender inequities that already result from oil wealth. Research by Ross shows that oil has a negative effect on the number of women in the labor force, which subsequently reduces their role and influence in politics.[3] Joel Simmons finds a similar effect in the United States.[4] Although more research into this issue is surely needed, we have little reason to be optimistic about the situation of female entrepreneurs in countries with abundant oil wealth.

A business environment that encourages firm owners to continually engage in bribery will bleed into other government–society interactions and create a culture of corruption, which continues to be a problem especially for states with oil.[5] In a direct sense, bribery exacts a substantial cost on firms,[6] which some research finds to adversely affect young firms and SMEs in particular.[7] Bribery can also create a culture of corruption in a society, which can have a broader effect on economic and political development. Although the empirical link between corruption and economic outcomes is complex, many view corruption as a symptom of a larger problem with the institutional foundations in a country. Susan Rose-Ackerman holds that being forced to engage in corrupt behavior by paying bribes is "always a second-best response to government failure."[8] Furthermore, bribery and corruption can erode the citizenry's level of trust in the government. High corruption and low trust in political institutions tend to coexist in a vicious cycle.[9] Although the relationship between trust and democracy is context specific and conditioned on the existing level of democracy,[10] fostering greater trust in political institutions can increase perceptions of "legitimacy" and encourage greater public commitment to governmental laws and regulations.[11] Bribery erodes this sense of legitimacy.

Finally, a restrictive business environment will weaken a country's SME sector and render these firms less capable of capitalizing on domestic opportunities for growth and investment. One of the biggest opportunities for SMEs is to serve as suppliers for larger firms. Fostering supplier relationships has a direct benefit for SMEs, the large firms who serve as customers, and consumers of large firms' products and services. Using domestic SMEs as suppliers can help solve some of the common problems that emerge from importing inputs, such as high transport or customs costs or issues with quality control. But if the SME sector is weak because of government-induced barriers, it is unlikely to be courted by large firms offering supplier contracts. This may result in a key missed opportunity for the transfer of new technology, better managerial practices, and investment capital from large firms to smaller firms. The writings on corporate venture capital—whereby a large firm obtains an equity stake in a smaller firm for strategic and/or financial reasons—outline the benefits that small firms experience when large firms invest in them, such as increased risk sharing, pooling of resources, and transfer of technical expertise.[12] A weak SME sector that is subject to excessive regulation is an unattractive site for corporate venture capital.

In Chapter 3, I offered a probabilistic theory of why an oil windfall results in a restrictive business environment. I outlined a "logic of policy making" in oil producers, something that is rarely done with respect to these countries. I argued that an oil windfall dampens the interests of policy makers and business elites about business environment reforms, and blunts the capacity of entrepreneurs and SMEs to demand these reforms. I argued that business elites have preferences about reforms that contrast with those of new entrepreneurs and SMEs. Whereas the latter group desires reforms that would lower barriers to entry, ease credit access, and streamline government regulation, business elites want to block these reforms as a means of limiting competition and raising their profits. Uncertainty about the outcomes of business environment reforms should increase their efforts to block them, given the potential for new entrepreneurs to emerge as a competitive threat down the road.

The oil windfall leads policy makers to be responsive to the interests of business elites and to form their own interests that are generally opposed to these reforms. This is the "supply-side" component of the theory. The decision to forego reforms can help solidify a policy maker's relationship with business elites, and they should be eager to do this if elites can serve as an important source of political strength. The high value that policy makers attach to satisfying the interests of business elites was vividly observed in the case study of Iran in Chapter 4. After the oil windfall and during the rule of Mohammad Reza Shah, business elites were his bastion of support, and he structured policy and economic institutions in order to benefit this group. After the 1979 revolution, *bonyads* and other units developed a similar relationship with the government that has been highly beneficial for both parties. *Bonyads* serve as a strong source of political support for the government—functioning as its proxy in some respects—and correspondingly they receive a wide range of benefits and lucrative business opportunities that other economic actors do not receive.

Policy makers allocate rents after an oil windfall in order to help them retain power and continue their command over an oil sector, which offers ancillary professional and material benefits. The broad range of benefits that policy makers receive because they control access to a resource windfall was illustrated in the case of Jharkhand in Chapter 5. Rent seeking led policy makers to grant coal and mineral leases to select business elites, as well as provide other forms of preferential treatment. But these were not gifts handed out as charity. They were exchanged for cash, lucrative positions at elite-owned firms, or political support given through donations to political parties. There was evidence of rent seeking and rent seizing at the *local* level in Jharkhand. Bureaucrats and public officials received bribes from the coal mafia in order to help facilitate the mafia's activities. Indeed, retaining control over the decision-making process that governs Jharkhand's mineral wealth has translated into political, professional, and material benefits for policy makers.

The constant practice of exchanging rents for benefits affects policy makers' cost-benefit calculus regarding the decision to adopt business environment reforms. We know that the economic benefits of reforms should be important to policy makers following an oil windfall, but these are likely to be significant only in the long term. In the short term, reforms impose economic costs and are a net fiscal loss to the state. Policy makers are likely to further discount the long-term economic benefits of reforms—but overestimate their short-term costs—if the stability of the government is in question. Policy makers should consider the political implications of avoiding reforms given that this would agitate nonelites, a numerically large segment of the private sector. Nevertheless, nonelites (by definition) are economically weaker at the individual level than elites. Their high degree of diversity should lead policy makers to perceive that they have considerable challenges in lobbying or engaging in collective action.

The "demand-side" component of the theory argues that the declining role of nonoil sectors on relative terms following an oil windfall affects the policy-making influence of nonelites. The government's increasing attention toward expanding the oil sector should lead nonelites to have less collective capacity to formally engage with policy makers. While nonelites already have little political capital at the individual level, we would expect the same at the collective level and with respect to formal channels of influence, such as through business associations. Nonelites may retain the capacity to engage in collective action to voice their demands, but this should lower their chances of getting a seat at the policy formation table.

Iran and Jharkhand provided evidence of these dynamics. After Iran received its oil windfall, the government squeezed the numerically dominant bazaar sectors on economic, political, and even physical terms. The late-Pahlavi era showcased a government intent on eliminating the bazaar, thus heightening tensions between the two. Yet, nonelites could not force major changes in the direction of policy. Business associations in Iran were either under the control of leaders, inherently weak, or in a state of opposition with the government. The case of Jharkhand highlights a number of similar dynamics. Despite the prevalence of active business associations throughout India, business associations are reportedly weak in Jharkhand and have little capacity to encourage policy makers to break from a mineral-led development strategy. Firms in nonmining sectors tend to mistrust one another and there is a lack of cooperation. Entrepreneurs often view the government as having failed at policy making that impacts their sector. Bihar, by contrast, has gone down the path of a more supportive business environment since losing its resource windfall to Jharkhand. Bihar has imposed fewer regulatory barriers on small-sized firms, provided more credit, and implemented more industrial reforms that benefit local entrepreneurs. There is evidence that Bihar's business associations directly engage with policy makers, and firm owners appear less frustrated with policy making and the current business environment.

Chapter 2 highlighted the broad range of outcomes that result for oil producers after these causal mechanisms are set in motion. There is strong evidence of a positive association between oil income and the degree of start-up regulation, tax regulation, and contracts regulation that domestic entrepreneurs and SMEs encounter. Higher levels of oil income are also associated with lower credit lending. Iran during the White Revolution was a prime example of the extent to which, in spite of having more governmental revenues available to devote to credit, political forces dictated and distorted the actual credit-lending process. Oil wealth also affects the day-to-day activities of existing firms. Managers of firms located in oil-producing countries report having to spend more time dealing with government regulation. They are more likely to pay bribes to officials to "get things done" with respect to regulation. Unsurprisingly, oil wealth is linked to lower rates of entrepreneurship and fewer patent applications by residents. These findings hold up after performing statistical tests that control for political, economic, and other factors. They hold up after conducting numerous robustness checks. I confirmed that the findings were not due to the inclusion of certain countries or world regions in the data set, such as the large number of Middle Eastern oil producers.

But oil is not the only resource that generates these outcomes. My controlled comparison of Bihar and Jharkhand illustrated that the production of coal yields similar rent-seeking and rent-seizing dynamics, a weakening of nonmining sectors, and policy-making outcomes that are akin to what emerges in oil producers. The presence of a democratic state did not shield Jharkhand from the resource windfall's effect on the contours of government policy making. The fact that policy making occurs at the subnational level in Jharkhand does not seem to matter. As long as a state government has autonomy in constructing its policies toward the local business community, as is the case in federal India, it is not immune from the distortionary effects of a resource windfall. To be sure, the book's findings about nonoil resource windfalls need to be investigated further. What is it about a natural resource that matters in terms of generating these outcomes? A common assumption in the literature seems to be that oil is more likely to generate negative outcomes simply because it is a high-value resource.[13] The higher the value of a resource on the international market the more distortionary it should be in terms of political, economic, and social outcomes. My comparison of Indian cases suggests that the *relative local value* of a resource is also important. In the context of Jharkhand, it is relevant that the state has a very low per capita income and the majority of nonmining economic opportunities exist in the low-value agricultural sector. Although Jharkhand classifies as mineral dependent by conventional standards, this number hides the fact that much of the rest of the state's economy is dominated by activities that yield markedly lower returns than mining or mining-dependent sectors. Furthermore, when considering the effect of coal at the local level in Jharkhand, it matters that coal is a lootable natural resource.

But geology does not entirely determine a state's destiny. Not all oil producers adopt policies that are detrimental to the rise of entrepreneurs, SMEs, competitive markets, and innovation. In Chapter 2, the Gulf Arab oil producers were observed to be exceptions to some of these trends, specifically in terms of business regulation and credit-lending practices over the last two decades. Chapter 6 focused on one of these cases, Saudi Arabia, a country that has adopted more policy reforms over the last fifteen years than the vast majority of its oil-producing colleagues. Saudi

Arabia has a stable monarchical system that reinforces the institution of economic familism, which is what I argue differentiates it from non-Gulf Arab oil producers who have a more restrictive business environment.[14] I argued that uncertainty about the outcomes of business environment reforms is reduced when there is a stable dynastic monarchy that continually legitimizes and reinforces the familial ties between policy makers and the private sector (as well as within the private sector). In Saudi Arabia, the monarchy's reinforcement of the institution of economic familism helps to provide a commitment to business elites that their rents will continue to be protected over the long run. Consequently, business environment reforms are unlikely to threaten their interests. Throughout the MENA region, monarchies are more likely than nonmonarchies to have adopted business environment reforms. In the *Doing Business* data set, seven of the ten highest-ranking business environments in the MENA region are in countries with monarchical rule. No monarchy in the MENA region has a ranking below 105 out of 185 countries.

Can the lessons from Saudi Arabia be applied to other contexts? Indeed, economic familism is an inherently culturally determined institution and the persistence of a monarchy is specific to certain historical processes. Only a few monarchies exist beyond the MENA region. But we can think about the role that political stability (measured as the risk of regime change) might play in terms of facilitating the adoption of business environment reforms. Consider what might follow if the risk of regime change is *high*. In this context, policy makers are more likely to utilize policy as a tool to help garner political stability and will quickly devalue the long-term economic benefits of reforms. A high risk of regime change should make policy makers less likely to adopt reforms that are perceived to antagonize business elites.[15] Basically, reforms should be more difficult. But any test of this hypothesis needs to first deal with the thorny issue of how an oil windfall itself affects political stability and what types of stability (regime durability, the absence of civil conflict or protest, etc.) actually matter most when it comes to facilitating policy reform.[16] In a different vein, we might think more about how economic familism operates as an *informal* institution that substitutes for a formal institution, a subject that is of increasing interest to scholars of comparative politics and political economy.[17]

The issue of political stability leads us to wonder about the political consequences of adopting business environment reforms in oil producers. In some sense, the political consequences of reforms seem obvious, but whether or not they materialize is difficult to predict. Adopting business environment reforms and providing more opportunities for individuals to seek economic opportunities and employment should reduce pressures on governments due to demographic bulges or economic crises. It is worth noting that urban, unemployed individuals were a highly mobilized group during the 1979 revolution in Iran, as they were in Tunisia and Egypt during the more recent events of the Arab Spring. Extending credit to new entrepreneurs and existing SMEs can be a way that a government seeks to bolster political support. A government that is perceived to be actively engaging in reforms and institution-building efforts that improve the business environment should be perceived by the citizenry as responsible "caretakers" of the economy.

The negative consequences that result from a restrictive business environment should be enough reason for policy makers in oil-producing countries to seek ways to pursue reforms—but there is another important reason. Simply put, management of an oil windfall is not an easy task. Ross addresses this issue in his discussion of the "fallacy of unobserved burdens":

> When oil-funded governments handle their revenues poorly, observers often blame it on the government's institutional weakness. This assumes that managing volatile resource revenues is no more burdensome than managing a much smoother flow of tax receipts. But maybe enacting consistent countercyclical policies is more onerous than we realize. Perhaps the problem is not that the oil states have exceptionally weak institutions and need normal ones; perhaps they already have normal institutions, but need exceptionally strong ones.[18]

We know that a number of challenges emerge when governments try to manage their oil revenues. We know that they, like all countries, have a limited capacity with which to realistically deal with these challenges. Just as oil is a finite resource, so are the human resources of any country. Governments that utilize the most talented bureaucrats to manage the country's oil sector are likely to send less talented ones to deal with the task of supervising manufacturing permits and filing credit applications from SMEs. This seems entirely understandable and expected. But as I have argued throughout this book, the policies and institutions that govern activities in these countries' nonoil sectors are just as important to address as what happens in their oil sectors. The nonoil sectors become even *more* important over time.

As long as governments task themselves with the job of managing an oil windfall, as they do in most countries around the world, there are two possible strategies for them to avoid the stagnation of their nonoil sectors.

One option is to spread the government's focus across the economy and invest in nonoil sectors that, unlike oil, are actually sustainable. Basically, this is where the government tries to do everything at once. This is the strategy that many oil producers have attempted at various times since the oil boom, and the historical record demonstrates that the results have been unimpressive and wasteful. The other option is to increasingly shift some of the responsibility for the growth of nonoil sectors onto others—either individuals who are based in the country or outside. For those seeking a more domestic-driven process, one way to achieve this is by actively pursuing the policy reforms discussed in this book: regulatory reforms, credit reforms, and pro-SME economic policies.

Beyond these reforms, there are other policies that can help counter the stagnation of nonoil sectors and low rates of innovation in oil producers. Governments that use public spending to foster an innovation infrastructure should generate higher innovation rates. Fewer patent applications may reflect weak linkages between the private sector and research institutes or universities. Researchers from universities and the private sector often collaborate and, in many fields such as pharmaceuticals and biotechnology, individuals move back and forth between the two. State-sponsored initiatives—such as increased support for research and development; the construction of technology parks; the provision of financial assistance to establish linkages between universities and the private sector; and the provision of subsidies to new firms operating in high-technology sectors—should all positively affect innovation in oil producers.

As time passes, there is an increasing sense of urgency for oil producers in the developing world to establish the foundations needed to diversify their economies. In recent decades, most of the oil producers who have sought diversification have done so by increasing trade, boosting tourism, or inviting foreign direct investment. Many of the business environment reforms we associate with the Gulf Arab oil states find complements in reforms of financial sectors and with respect to investment. But not all oil producers can be like Dubai or the other successful cases located in the Gulf Arab region, as these places possess certain advantageous characteristics that are wholly absent in most other contexts. And perhaps not all oil producers should *want* to become places like Dubai in light of the banking and real estate crises that hit the city in the late 2000s. As of writing, there are still many unanswered questions about the long-term risks and sustainability of the "Dubai model."[19]

There are fewer risks in fostering the domestic foundations for private-sector growth to emerge in a more organic process over time. In order to do so, oil producers will need to address regulation, credit, and barriers to entrepreneurship and the activities of small- and medium-sized firms. They will need to invest in human capital, provide training for entrepreneurs, and commit themselves to using policy as a tool to support the growth of nonoil sectors of the economy. Oil producers may have failed to capture the opportunity that oil brought them during the twentieth century, but they can take steps to ensure that this same opportunity is not lost during the twenty-first century.

APPENDIX

List of Interviews Cited in Chapter 5

Adukia, Himanshu, attorney, interview, Ranchi, Jharkhand, July 18, 2012.

Agarwal, Arun, president of Bihar Industries Association, interview, Patna, Bihar, January 24, 2014.

Ahuja, Rahul, director of manufacturing firm, interview, Ranchi, Jharkhand, July 27, 2012.

Baiju, owner of SME in wood furniture sector, interview, Patna, Bihar, January 28, 2014.

Barman, Nirmal, journalist, interview, New Delhi, July 10, 2012.

Bhushan, Chandra, deputy director general of Centre for Science and Environment, interview, New Delhi, January 3, 2014.

Chand, M. B., owner of SME in wood furniture sector, interview, Patna, Bihar, January 28, 2014.

Chowdhury, Samik, economist and professor at Institute of Economic Growth, interview, New Delhi, December 26, 2013.

Dayal, Harishwar, economist, interview, Ranchi, Jharkhand, July 26, 2012.

Dinu, owner of SME in wood furniture sector, interview, Ranchi, Jharkhand, July 20, 2012.

Diwakar, D. M., economist and director of A. N. Sinha Institute of Social Studies, interview, Patna, Bihar, January 21, 2014.

Gulab, Ram, manager at Jharkhand Institute of Entrepreneurship, interview, Ranchi, Jharkhand, July 19, 2012.

Gupta, Maganlal, owner of SME in wood furniture sector, interview, Patna, Bihar, January 29, 2014.

Gupta, Shaibal, member-secretary of Asian Development Research Institute, interview, Patna, Bihar, January 14, 2014.

Jashnuppal, owner of SME in wood furniture sector, interview, Ranchi, Jharkhand, July 20, 2012.

Khan, Mohammed, Jharkhand researcher and activist, interview, New Delhi, January 8, 2014.

Khemka, Arun Kumar, president of Jharkhand Small Industries Association, interview, Ranchi, Jharkhand, July 3, 2012.

Kumar, Arvind, owner of SME in wood furniture sector, interview, Patna, Bihar, January 29, 2014.

Kumar, Ashok, owner of SME in wood furniture sector, interview, Patna, Bihar, January 29, 2014.

Kumar, Dilip, executive at Confederation of Indian Industry, interview, Patna, Bihar, January 14, 2014.

Kumar, Lalit, chartered accountant, interview, Patna, Bihar, January 23, 2014.

Kumar, Sanjay, director at the Centre for the Study of Developing Societies, interview, New Delhi, July 16, 2012.

Kumar, Sushil, Jharkhand researcher, interview, New Delhi, January 9, 2014.

Lal, Pyare, director at Bihar Institute of Economic Studies, interview, Patna, Bihar, January 27, 2014.

Mitra, Ashim, owner of manufacturing firm, interview, Patna, Bihar, January 27, 2014.

Nandy, owner of SME in wood furniture sector, interview, Ranchi, Jharkhand, July 21, 2012.

Pandey, Rohit, owner of manufacturing firm, interview, Ranchi, Jharkhand, July 27, 2012.

Prakash, Amit, political scientist at Jawaharlal Nehru University, New Delhi, January 5, 2014.

Prasad, Ranjan and Vasudev, owners of SME in wood furniture sector, interview, Patna, Bihar, January 28, 2014.

Punjabi, Rupesh, owner of SME in wood furniture sector, interview, Ranchi, Jharkhand, July 20, 2012.

Ravi, owner of SME in wood furniture sector, interview, Patna, Bihar, January 29, 2014.

Ray, Sweta, Jharkhand researcher, interview, New Delhi, February 10, 2014.

Roy, Satyaki, economist at Institute for Studies in Industrial Development, interview, New Delhi, January 9, 2014.

Sharan, Ramesh, economist, interview, Ranchi, Jharkhand, July 26, 2012.

Sharma, Mahendar, owner of SME in wood furniture sector, interview, Patna, Bihar, January 28, 2014.

Sharma, S., owner of SME in wood furniture sector, interview, Ranchi, Jharkhand, July 20, 2012.

Sharma, Vikas, owner of SME in wood furniture sector, interview, Ranchi, Jharkhand, July 21, 2012.

Shekhar, Mayank, attorney, interview, Patna, Bihar, January 15, 2014.

Siddiqui, M. L., management consultant, interview, Ranchi, Jharkhand, July 4, 2012.

Singh, Awadhesh, attorney, interview, Patna, Bihar, January 16, 2014.

Singh, Deen Bandhu, attorney, interview, Patna, Bihar, January 23, 2014.

Singh, Kanhaiya, economist at National Council of Applied Economics Research, interview, New Delhi, July 11, 2012.

Singh, Rahul, head of Jharkhand state office of Confederation of Indian Industry, interview, Ranchi, Jharkhand, July 23, 2012.

Singh, R. D., secretary general of Federation of Jharkhand Chamber of Commerce and Industries, interview, Ranchi, Jharkhand, July 18, 2012.

Singh, R. U., former principal secretary to chief minister of Bihar, interview, Patna, Bihar, January 27, 2014.

Singh, Satyajit Kumar, chairman of the PHD Chamber of Commerce, interview, Patna, Bihar, January 24, 2014.

Sinha, Ajay Kumar, owner of SME in wood furniture sector, interview, Patna, Bihar, January 29, 2014.

Sinha, Indrajit, advocate at Jharkhand High Court, interview, Ranchi, Jharkhand, July 19, 2012.

Sinha, S., owner of SME in wood furniture sector, interview, Ranchi, Jharkhand, July 21, 2012.

Srivastava, Bijay, owner of SME in wood furniture sector, interview, Patna, Bihar, January 28, 2014.

Srivastava, K. K., mining sector consultant, interview, Ranchi, Jharkhand, July 23, 2012.

Verma, Dinesh Kumar, owner of SME in wood furniture sector, interview, Patna, Bihar, January 28, 2014.

NOTES

CHAPTER 1

1. Bozorgmehr 2009.
2. Forbes 2007.
3. Arkhipov 2010.
4. See Ross 2001a.
5. I use the word *oil* to refer to both oil and natural gas. I use the terms *oil wealth*, *oil income*, and *oil and gas income* interchangeably. The term *nonoil* refers to all nonpetroleum natural resources of a mineral or nonmineral nature.
6. Ross finds that oil states experience more economic volatility than nonoil states but their economic growth rates are no different. That said, "The real mystery is why the oil states have had normal growth rates, when they should have had faster than normal economic growth, given their enormous natural wealth" (2012, 189).
7. I use the term *Gulf Arab oil state* to refer to the monarchical oil states in the Gulf region: Bahrain, Kuwait, Oman, Qatar, Saudi Arabia, and the United Arab Emirates.
8. I use the terms *private sector* and *business* interchangeably to refer to privately-owned firms, entrepreneurs, and investors.
9. Beck, Demirgüç-Kunt, and Levine 2003.
10. Audretsch, Keilback, and Lehmann 2006.
11. Klapper, Laeven, and Rajan 2006.
12. I use the terms *developing world* and *developing country* to refer to the approximately 150 countries located in Africa; Asia, and the Pacific region; the Caribbean; Eastern Europe; Latin America; and the Middle East, who are not members of the OECD. While there is no generally accepted way of classifying a country as "developing," OECD membership is often used by international organizations as a way of separating "advanced economies" from "developing economies." See the discussion in the International Monetary Fund working paper by Nielsen 2011.
13. Note that there is no universal definition of an "SME." The European Union defines SMEs as having between 10 and 249 employees and annual sales under twenty-seven million euros. Until recently, Iran defined SMEs as establishments with less than ten employees. India defines SMEs based on the value of investment, not according to the number of employees.
14. With the exception of Chapter 5, I am not concerned with firms that engage in natural resource extraction.

15. The "informal sector" is traditionally understood as a "heterogeneous group of activities and employment relationships that share one common characteristic—the lack of legal recognition, regulation or protection" (Lloyd-Evans 2008, 1885).
16. This is often termed the "semi-formal" economy. See Kamrava 2004.
17. Thurik and Wennekers 2004.
18. International Finance Corporation 2010, 6. See Ayyagari, Demirgüç-Kunt, and Maksimovic 2011, 2, and Ayyagari, Beck, and Demirgüç-Kunt 2005.
19. Mourougane 2012.
20. See http://www.sbecouncil.org/about-us/facts-and-data/#sthash.cJHSF2SO.dpuf. Accessed September 3, 2014.
21. Refer to Levine 2005 for an overview of "pro-SME" and "skeptical-SME" arguments and findings.
22. Aghion and Howitt 1998; Baumol 2002; Wennekers and Thurik 1999.
23. Schumpeter [1911] 1934.
24. See Acs, Carlsson, and Karlsson 2011.
25. Szirmai, Naudé, and Goedhuys 2011; Dias and McDermott 2006; Narayan 2002.
26. See Tybout 2000; Dasgupta 2010; Krueger 2014; Hsieh and Olken 2014. Not all developing economies exhibit evidence of a "missing middle." E.g., Hsieh and Olken 2014 do not find evidence of such a phenomenon in India, Indonesia, and Mexico. See the discussion in Tybout 2014.
27. Djankov, McLiesh, and Shleifer 2007.
28. International Labour Organization 2002.
29. Schneider, Buehn, and Montenegro 2010. These figures are for 1997 to 2007.
30. De Soto 1989.
31. See the review and discussion in La Porta and Shleifer 2014.
32. See Bruhn and McKenzie 2013.
33. Refer to Ross 2001a and Haber and Menaldo 2011a.
34. See Moore 1966; Therborn 1977; Luebbert 1991; Rueschemeyer, Stephens, and Stephens 1992; Hsiao and Koo 1997; Norris 1999; Nasr 2009.
35. Rueschemeyer, Stephens, and Stephens 1992.
36. Bellin 2002.
37. E.g., refer to Chen 2002 on China, Shadlen 2002 on Mexico, and Greenwood 2008 on Jordan and Morocco.
38. Ross 2001a; Ross 2009.
39. On the political "resource curse" see Dunning 2008; Haber and Menaldo 2011a; Ross 2001a; Ross and Andersen 2014. Although the economic "resource curse" finding was taken as an article of faith for many years, researchers today question an unconditional effect of oil wealth on economic outcomes. See Alexeev and Conrad 2009; Lederman and Maloney 2007; Brunnschweiler and Bulte 2008; Wright and Czelusta 2002; Stijns 2005. Many demonstrate that intervening variables—such as the role of resource ownership (Jones Luong and Weinthal 2010), political institutions (Mehlum, Moene, and Torvik 2006; Robinson, Torvik, and Verdier 2006), the incentives of elites to promote economic diversification (Dunning 2005), geography (Wibbels and Goldberg 2007), political fractionalization (Bjorvatn, Farzanegan, and Schneider 2012), or the level of human capital (Bravo-Ortega and de Gregorio 2007)—are central to any emergence of an economic resource curse.
40. Beblawi and Luciani 1987; Mahdavy 1970. See the overview in Ross 2001a, 332–335.

41. Karl 2004, 666–668.
42. Smith 2007.
43. Chaudhry 1997.
44. In fact, the rentier state argument and related writings offer contrasting predictions about how oil wealth will impact political stability. See Morrison 2012. Smith 2004 finds that oil wealth increases the durability of regimes, decreases the likelihood of civil war, and decreases the likelihood of antistate protest.
45. On the link between political stability and economic reform see Bunce 2001; Haggard and Kauffman 1992; Van De Walle 1994.
46. Deacon and Rode 2012; Karl 1997.
47. Corden and Neary 1982; Auty 2001; Sachs 1999; Ploeg 2011.
48. Fardmanesh 1991; Gelb et al. 1988; Spatafora and Warner 1995; Jahan-Parvar and Mohammadi 2009, 2011; Sala-i-Martin and Subramanian 2003.
49. E.g., in Iran see Shams Ardekani 2013 and Nosrati 2013.
50. Sachs 2007, 184.
51. Karl 1997, 5. See also Poteete 2009.
52. Lowi 2009, 31.
53. The Dutch Disease is found to have a larger effect on the activities of exporting firms than nonexporting firms (Harding and Venables 2013). Nonexporting firms generally represent a larger share of total firms in any country and are the focus of this book.
54. Baland and Francois 2000; Bjorvatn and Selvik 2008; Papyrakis and Gerlagh 2005; Falkinger and Grossmann 2005; Mehlum, Moene, and Torvik 2006; Torvik 2002.
55. See Acemoglu and Robinson 2012; Amsden, DiCaprio, and Robinson 2012.
56. Barma et al 2012; Caselli and Cunningham 2009; Dunning 2005; Hausmann and Rigobon 2003; Humphreys, Sachs, and Stiglitz 2007; Kolstad and Wiig 2009; Ross 2001b; Ross 2012, ch. 7; Weyland 2009.
57. See Frankel 2010; Humphreys, Sachs, and Stiglitz 2007; Weinthal and Jones Luong 2006.
58. Merriam-Webster dictionary defines a *windfall* as: "an unexpected amount of money that you get as a gift, prize, etc."
59. On rent-seeking dynamics between government and business elites in oil states see Chaudhry 1997; Crystal 1995; Moore 2004; Okruhlik 1999; and Shambayati 1994. On rent seizing, see Ross 2001b.
60. Hallward-Driemeier and Pritchett 2011.
61. Hirschman 1958.
62. Karl 1997, 190.
63. Auty 2001, 844.
64. See Hvidt 2011 and the discussion in Moore 2002, 38.
65. The goal of process tracing is "to *connect* the phases of the policy process and enable the investigator to identify the reasons for the emergence of a particular decision through the dynamic of events" (Tarrow 1995, 472; see George and McKeown 1985). Using process tracing requires analyzing events over time by "taking good snapshots at a series of specific moments" and explaining the steps in the process that uncovers change. Constructing such a description often involves relying on quantitative data (Collier 2011, 824–825).
66. Collier 2010, 37.

CHAPTER 2

1. The metaphor extends to sports. A 2013 article in *ESPN* quoted the work of Terry Lynn Karl and reasoned that the success or failure of baseball teams can be explained through the resource curse argument (Miller 2013).
2. Bodin [1576] 1967; Smith [1776] 1976.
3. For an overview, see Frankel 2010.
4. GDP data were obtained from Haber and Menaldo 2011b, which I update using data from World Bank 2015 and the authors' approach (see Haber and Menaldo's "Appendix on Sources and Methods," p. 28). Oil and gas production data were obtained from Ross 2013. I do not use export concentration or GDP-based measures of oil wealth given that both of these measures suffer from bias (Alexeev and Conrad 2009, 589; Ross 2012, 15–17). Instead, I adopt continuous and binary measures of oil abundance (oil income per capita), per Ross 2012.
5. Like all quantitative measures this one has limitations. It is unable to distinguish between nonoil income that is generated from the public vs. the private sector, the local- vs. joint local/foreign-owned private sector, or reveal other nuanced aspects that are important to parse out in the case studies that appear later. Nevertheless, it is an appropriate and, more importantly, easily comparable estimate of everything but oil in an economy.
6. The data set is 95 percent complete (after adjusting for countries that became independent after 1960).
7. Ross 2012, 16 and 20–22.
8. Crude oil prices are in real September 2013 dollars and taken from the US Energy Information Administration website. See http://www.eia.gov/petroleum/data.cfm#prices.
9. The data set is available at http://www.doingbusiness.org. While I additionally rely on measures from another data set, the World Bank's *Enterprise Surveys* (see http://www.enterprisesurveys.org), the *Doing Business* data set has a broader coverage of countries (40 percent more countries sampled), many of which are oil producers in the developing world. And unlike the *Enterprise Surveys*, the *Doing Business* data set is focused almost entirely on the issue of government regulation and entrepreneurship. Although the *Doing Business* data set has been widely used in political economy studies since its introduction, it is sometimes criticized for potential bias in its measures (e.g., see Davis and Kruse 2007; see also note 32 in the following text). For a defense of the data set and discussion of these criticisms see Acemoglu et al. (2013). To be sure, relying on a single data source or indicator will provide a limited insight into the nature of a country's business environment. For this reason, I rely on a variety of data sources and examine a range of indicators in the large-N analysis of this chapter. In the subsequent case study chapters, I adopt a holistic approach to understanding the business environment in each context.
10. See Djankov et al. 2002 for details. Some measures employ different definitions of a firm and I note when this occurs. This becomes important when assessing the India cases in Chapter 5.
11. Ibid.
12. The data set's measure of bureaucratic time delay is "the median duration that incorporation lawyers or notaries indicate is necessary in practice to complete a procedure with minimum follow-up with government agencies and no unofficial payments" (http://www.doingbusiness.org/methodology/starting-a-business).

Among developing countries, the correlation between the number of start-up procedures and the time to complete a procedure is 0.385.

13. Among developing countries, the correlation between permit procedures and their time delay is 0.118.
14. Among developing countries, the correlation between tax payments and their time delay is 0.243.
15. Among developing countries, the correlation between contract procedures and their time delay is 0.257.
16. Klapper, Laeven, and Rajan 2006.
17. Friedman et al. 2000.
18. Braunerhjelm and Eklund 2013; see also Djankov et al. 2010.
19. Rodrik 2008, 3–5. See the review in Haggard, MacIntyre, and Tiede 2008.
20. Djankov, McLiesh, and Shleifer 2007.
21. World Bank 2013.
22. E.g., see Duflo et al. 2013.
23. Beck 2007.
24. Galindo and Micco 2007.
25. See Hausmann and Rigobon 2003; Gelb and Grasmann 2010.
26. Data are from World Bank 2015.
27. Data are from ibid. It should be noted that this measure, along with *Depth of Credit Information*, applies to firms that are somewhat larger in size (one hundred employees) than those that are the focus of this book.
28. This is when controlling for relevant confounding variables and including country clusters, time dummies, region dummies, and robust standard errors.
29. See Achen 2002 and its application in Ross 2012.
30. Djankov et al. 2002.
31. Data for *regime type* and *regime durable* are taken from the *Polity* data set. See Marshall and Jaggers 2012.
32. Prior to conducting statistical tests, all business regulation measures are log transformed in order to reduce skew. A working paper from the World Bank's Enterprise Analysis Unit looks at the association between primary commodity exports and a count (or dummy) measure of the reforms a country has undertaken in the *Doing Business* data over the time period 2003 to 2008 (Amin and Djankov 2009). However, a count of the number of reforms over a highly truncated period does not tell us how primary commodity producers compare overall to nonproducers—only the number of reforms that occurred during a five-year period. There is evidence that governments have incentives to try and manipulate their scores on the *Doing Business* measures over time due to the signal that it may send to international investors. A World Bank employee told me that he was approached by government officials from two separate countries following the release of the *Doing Business* data set for advice on how to "game the indicators." A separate encounter echoed this impression. For this reason, comparing where countries are situated on these measures overall—and not the number of reforms during the span of data collection—is a better way to examine differences between oil producers and nonproducers. Additionally, the authors employ GDP- and export-dependence measures of primary commodities that can produce misleading statistical results (Alexeev and Conrad 2009). A separate World Bank study examines the challenges of diversification in a group of MENA oil exporters (Diop, Marotta, and de Melo 2012; see also Gelb 2011), but topics such as entrepreneurship, SMEs, and credit are not addressed.

33. Clustering by country is always done for the statistical tests on the business regulation measures. The intraclass correlations for country-clustered observations are 0.7 for *start-up regulation*, 0.92 for *permits regulation*, 0.91 for *tax regulation*, and 0.98 for *contracts regulation*. This suggests relatively little change in a country's values during the time period of analysis.
34. The statistical association holds when the regression is performed on the full sample of all countries in the world. This is expected given that the majority of OECD countries have both low levels of oil income and low levels of start-up regulation, as the theory would predict.
35. Karl 1997, xv.
36. Gelman 2007.
37. Fisman and Svensson 2007; see also Shleifer and Vishny 1993.
38. Data are from World Bank 2015.
39. Klapper and Love 2010; see also Van Stel, Storey, and Thurik 2007.
40. The intraclass correlation is 0.89 for country-clustered observations.
41. In the context of Latin America, Maloney 2007 holds that a negative relationship exists between natural resources and innovation. Blomström and Kokko 2007 examine the Scandinavian context to show how the natural resource sector pushes innovation and the development of high technology. Papyrakis and Gerlagh 2005 run bivariate regressions on small cross-sectional data sets (containing thirty-one to fifty-two countries) and find that natural resources are negatively linked to research and development spending and the number of patents granted.
42. Data are from World Bank 2015. Bottazzi and Peri state that "patents can be considered as a good approximation to ideal data on 'economically profitable ideas' which one would like to have for testing theories on innovation" (2003, 692).
43. Heller and Eisenberg 1998, 699. For this study, a measure of patent applications is a better indicator of innovation than alternatives such as a country's high-technology exports, which is skewed by the presence of foreign firms.
44. Patent data stretches back to 1963 for twenty-four developing countries and the data set is 48 percent complete as a whole.
45. I multiply this measure by one million for ease of interpretation.
46. The difference in means between long-term oil producers and nonproducers is significant at the same level.
47. Barro 2001; Cipolla 1969.
48. Data are from World Factbook 2014 and World Bank 2015.
49. Data are from World Bank 2015. Controlling for foreign direct investment (FDI) in the regression analysis invites post-treatment bias because the presence of oil wealth leads to more FDI.
50. Substituting a measure of patents per capita and removing total population as a control does not alter the results.

CHAPTER 3

1. Huntington 1952.
2. Breyer 1982; Slinko, Yakovlev, and Zhuravskaya 2005; Stigler 1971.
3. Haggard and Kaufman 1992.
4. Bates 1981.
5. Grindle 2005, 1.
6. Krueger 1974.
7. Djankov et al. 2002; De Soto 1989; Shleifer and Vishny 1994.
8. Olson 1965; Stigler 1974.

9. Weymouth 2012.
10. Kerr, Lincoln, and Mishra 2011.
11. Macher, Mayo, and Schiffer 2011, 21.
12. Doner and Schneider 2000; Evans 1995; Maxfield and Schneider 1997.
13. Calì and Sen 2011.
14. See Olson 1982.
15. Acemoglu 2008; Acemoglu and Robinson 2012.
16. That said, one can find examples of "super-elites" voicing support for the economic interests of nonelites. Although it is unclear how sincere this support is, they are usually figures in the public eye, such as Mikhail Khodorkovsky in Russia or Prince Al-Waleed bin Talal in Saudi Arabia. While the case of Khodorkovsky is likely influenced by his personal experiences with the Russian state, for these select few individuals, nonelites pose no serious threat to their business interests and this might help facilitate a more reform-minded perspective.
17. Hallward-Driemeier and Pritchett 2011; Hellman and Kaufmann 2003; Slinko, Yakovlev, and Zhuravskaya 2005; Stone, Levy, and Paredes 1992.
18. See Crystal 1995 for more details and analysis.
19. Ibid., 10.
20. Chaudhry 1997.
21. Thorp and Durand 1997, 230–231.
22. Arrow et al. 1996; Government of Australia 2007; Sunstein 2003; Treasury Board of Canada 2007.
23. See the discussion in Pearce 1997.
24. See Livermore and Revesz 2013.
25. Menaldo 2016.
26. Wildavsky 1966.
27. Ross 2001b.
28. Chaudhry 1997.
29. Grindle 2005, 9–10.
30. Historical forces, government-elite relations, and the role of networks play important roles in shaping the policy influence of business associations. See Cammett 2007; Frye 2002; Haddad 2012; Moore 2001, 2002; Peregudov and Semenenko 1996; Schneider 1997.
31. Moore and Salloukh 2007.
32. Ibid., 57.
33. Moore 2001. Jordan is sometimes thought of as having many of the features of a rentier state given its dependence on foreign aid instead of oil. Some research shows that aid and oil have similar political effects (Morrison 2009), but there is also contrasting evidence (Bermeo 2016). Collier (2006) discusses the different developmental effects of aid versus oil.
34. See Doner and Schneider 2000; Maxfield and Schneider 1997; Sinha 2005.
35. See also the analysis of Syria in Moore and Salloukh 2007.
36. See the discussion in Evans and Stephens 1988.
37. See Parsons 1951; Huntington 1968.
38. See Scott 1977.

CHAPTER 4

1. Katouzian 1981, 204.
2. Historical oil production data are from Haber and Menaldo 2011b.
3. George and Bennett 2004, 166–167.

4. Individuals in this group are referred to as *bazaaris*.
5. Moaddel 1986, 541; Ghandchi-Tehrani 1982, 43.
6. Bashiriyeh 1984, 103; Keshavarzian 2007, 6.
7. In the city of Qazvin, the bazaar controlled up to 60 percent of economic activities. By the mid-1970s, the Qazvin bazaar was critical to the livelihoods of one-third of all Iranians (Smith 2007, 116; Denoeux 1993, 139; see also Rotblat 1972). Some bazaars in Iran, such as the one based in the city of Tabriz, have existed since the days of the Silk Road.
8. Bayat 1991, 45; Keddie 1980, 209–210.
9. Keddie 1980, 130.
10. Ibid., 210.
11. Katouzian 2009, 221.
12. Amirahmadi 1986, 503–504.
13. Bayat 1991, 45.
14. Paine and Schoenberger 1975, 3–4.
15. Keddie and Amanat 1991, 181.
16. Moaddel 1986, 528.
17. Keddie 1983, 583.
18. Keddie 1980, 130–131.
19. See Abrahamian 1979, 393; Ghandchi-Tehrani 1982, 99–100.
20. Moaddel 1986, 529–530.
21. Issawi 1978, 131.
22. Katouzian 1981, 95.
23. Keddie 1983, 592.
24. Wilber 1948, 302; Amirahmadi 1986, 503–504; 511.
25. Mehner 1978, 197.
26. Wilber 1948, 303.
27. Ibid., 304.
28. Katouzian 2009, 221.
29. Abrahamian 1982, 151–152. See also Yaghoubian 2014, ch. 3.
30. Keddie 1980, 135; Zonis 1971, 320.
31. See Abrahamian 2013 for more details and analysis.
32. Karshenas 1990, 110.
33. Parsa 1994, 138.
34. Ganji 2002, xvii.
35. MacLachlan 1991, 615. See Karshenas 1990 for more on the economic situation in Iran during the 1950s.
36. Parsa 1994, 144–145; Ramazani 1974, 131; Keddie 1983, 588.
37. Foran 1994, 167–168.
38. Ibid., 167.
39. Zahedi 2000, 94–95.
40. Abrahamian 1982, 430; see also Issawi 1978, 150.
41. See Shambayati 1994.
42. Zonis 1971, 320–321.
43. Ghandchi-Tehrani 1982, 85.
44. Amirahmadi 1986, 514.
45. Keddie 1980, 220–221.
46. Baldwin 1967; quoted in Carey and Carey 1975, 7.
47. Ghorayshi 1981, 28–29.
48. Salehi-Isfahani 1989.

49. Karshenas 1990, 105.
50. Issawi 1978, 155.
51. Mehner 1978, 195–196.
52. See Keshavarzian 2007, 87–89.
53. Ibid., 85.
54. Ibid.; see also Salehi-Isfahani 1989.
55. Keshavarzian 2007, 86.
56. Parsa 1988, 62.
57. Data for Iran's oil income are from Ross 2013. Credit data are from World Bank 2015.
58. Issawi 1978, 154.
59. Carey and Carey 1975, 4.
60. Issawi 1978, 154.
61. Zahedi 2000, 94.
62. Bashiriyeh 1984, 41; see also Issawi 1978, 155.
63. Data on new manufacturing permits are available from *Vezarat-e Sanaye va Maaden-e Iran* (Ministry of Industry and Mines) and *Bank Markazi Jomhouri-ye Islami-ye Iran* (Central Bank of the Islamic Republic of Iran) and available from Government of the Islamic Republic of Iran 2013a. Note that the ministry was reorganized as the Ministry of Industry, Mines, and Trade. Data for Iran's oil income are from Ross 2013.
64. Karshenas and Pesaran 1995, 95–96; Parsa 1989, 84.
65. Liu 1988, 190.
66. Shambayati 1994, 321; see also Keddie 1980, 220; Keddie 1983, 593.
67. Ashraf 1995, 34.
68. Ibid.
69. Nasr 2000, 111.
70. Parsa 1994, 155.
71. Nasr 2000, 113.
72. Moaddel 1986, 541.
73. Ghandchi-Tehrani 1982, 84–85.
74. Denoeux 1993, 145; Shambayati 1994, 321.
75. Ghods 1989, 200.
76. See Abrahamian 1978, 4; 1982, 497; Salehi-Isfahani 1989, 364.
77. Amuzegar 1991, 246; Abrahamian 1982, 498.
78. Zahedi 2000, 87; Mazaheri 2006.
79. Abrahamian 1982, 498.
80. Ibid., 440.
81. Amjad 1989, 99.
82. Foran 1993, 335.
83. Amuzegar 1991, 276.
84. Amjad 1989, 100.
85. Bashiriyeh 1984, 103; Zahedi 2000, 87.
86. See Graham 1980, 134.
87. Zahedi 2000, 87; Abrahamian 1982, 498.
88. Parsa 1989, 103–104.
89. Graham 1980, 96.
90. Ashraf 1995, 34–35.
91. Mozaffari 1991, 383.
92. Keshavarzian 2007, 242.

93. Bonine 1979, 210.
94. Keshavarzian 2007, 91.
95. Parsa 1988, 54.
96. Data are from Banks and Wilson 2013.
97. Dorraj 1990, 172; see also Ghandchi-Tehrani 1982, 12–17.
98. Pahlavi 1980, 156.
99. Mozaffari 1991, 379–380.
100. Karshenas 1990, 105.
101. This is witnessed, e.g., through statements by Ayatollah Khomeini and Ayatollah Beheshti that referred to "bloodsucking capitalists" (*sarmayehdaran-e zalusefat*). See Behdad 2000, 31–32.
102. Selvik 2005.
103. Saeidi 2004.
104. See Abrahamian 1993, 53–54.
105. *Sazman-e Basij-e Mostazafin* (or "Basij") is a paramilitary organization that is subordinate to *Sepah-e Pasdaran-e Enqelab-e Islami* (the Islamic Revolutionary Guards Corps, or IRGC) and was established after the 1979 revolution.
106. Resalat 2013.
107. Abrahamian 1993, 42.
108. Hunter 1992, 64.
109. Amuzegar 1992, 423; Amuzegar 2000; Clawson 1988; Ehsani 1994.
110. Ghasimi 1992, 602.
111. Rahnema 1995, 129.
112. This is evident from examining data on gross fixed capital formation. Data are from *Markaz-e Amar-e Iran* (Government of the Islamic Republic of Iran 2013b).
113. Dabashi 2007, 187–189.
114. Maleki 2009, 15.
115. Behdad 2001.
116. Maleki 2009, 15. See also Pesaran 2011.
117. Khajehpour 2007, 9.
118. Maleki 2009, 15.
119. Data are from World Bank 2015.
120. Etemaad 2013.
121. Jenanseft 2013.
122. Harris 2013b.
123. This is an aspect of *bonyads* that is often ignored by Iran commentators who focus only on the economic—but not social—role they play in the country. See Salehi Esfahani 2005.
124. Keshavarzian 2007, 167. See also Askari 2004, 661.
125. See Alizadeh 2003; Behdad 2000; Karshenas and Pesaran 1995.
126. Rahnema 1995, 138.
127. Farzin 1995.
128. Saeidi 2004.
129. Karshenas and Hakimian 2008, 201.
130. Behdad 2000, 27.
131. Farzin 1995, 183.
132. Behdad 2000, 27.
133. Farzin 1995, 177.
134. Bjorvatn and Selvik 2008, 4.

135. Askari 2004, 660; Farzin 1995, 177.
136. Saeidi 2004, 484.
137. Ibid., 486.
138. Amuzegar 1999, 548. *Bonyads* are often criticized for their economic inefficiency, but it is difficult to determine if this is a general pattern. Some argue that "the lack of technical and managerial experience, institutional rivalry, and corruption seriously [undermine] their effectiveness" (Hunter 1992, 64–65).
139. Bjorvatn and Selvik 2008, 2318; Yasin 2007. Iran observers often describe the economic activities of the IRGC as a growing "militarization" of the economy. This is further evidenced by the increasing turn toward economic activities among paramilitary groups, such as the Basij (Golkar 2012). However, this process may be less dramatic than the media tends to portray (Harris 2013b).
140. The OSF was later replaced by a National Development Fund.
141. Naghshineh-Pour 2008, 2.
142. See http://www.al-monitor.com/pulse/originals/2013/10/hope-for-iran-private-sector-national-development-fund.html#.
143. Heuty 2012, 1.
144. This is subject to certain exceptions, such as the national oil company.
145. Mir-Motahari 2013.
146. *Jaam-e Jam*. December 17, 2012.
147. *Jaam-e Jam*. December 31, 2012.
148. Hamshahri 2009.
149. Alef 2012e.
150. Etemaad 2009.
151. Safari Esfahani 2009.
152. Kamalian et al. 2011, 79.
153. Amuzegar 1999, 548.
154. Karimi et al. 2010, 43.
155. Zangeneh 2003, 108; see also Kian-Thiebaut 2001.
156. Tax data are compiled and available from *Markaz-e Amar-e Iran* (Government of the Islamic Republic of Iran 2013b).
157. Examples of indirect taxes and charges for firms are sales, use, customs, and excise taxes.
158. World Bank 2015.
159. See PricewaterhouseCoopers 2014, 110–117. On the subject of taxation, I am grateful for insights provided by David Cropper
160. Zangeneh 2003, 128.
161. Askari 2004, 663.
162. See Behdad 2000, 11–12; see also Ehteshami and Zweiri 2007, 74–76.
163. See Jalali-Naini 2008.
164. Eeles 2004.
165. Kamalian et al. 2011, 84.
166. See Keshavarzian 2007.
167. Abrahamian 1993, 143.
168. Hamshahri 2011.
169. Yong 2010.
170. Earlier that year, Seyyed Shamseddin Hosseini, Minister of Economic Affairs and Finance, spoke out against individual manipulators as causing currency fluctuations (*Jaam-e Jam*. May 17, 2012). See also Harris 2013a.
171. Gladstone 2012.

172. Alef 2012a.
173. Alef 2012b.
174. Kayhan 2012.
175. Alef 2012c. A number of bazaaris called for cooperation with the government to address these issues. See Hamshahri 2012.
176. Alef 2012d.
177. Press TV 2009.
178. Shargh 2011.
179. See Farhi 2013.
180. The following figures on manufacturing and trade employment are based on the author's calculations using data from Iran Census Centre and reported in Beheshti 2003. Literacy data are from *Markaz-e Amar-e Iran* (Government of the Islamic Republic of Iran 2013b) and reported in Beheshti 2003. Manufacturing data are compiled and available from *Markaz-e Amar-e Iran* (Government of the Islamic Republic of Iran 2013b).
181. See Behdad and Nomani 2009.
182. Gross national product data are reported in the National Accounts of Iran, calculated by *Bank Markazi Jomhouri-ye Islami-ye Iran* (Central Bank of the Islamic Republic of Iran) and available from *Markaz-e Amar-e Iran* (Government of the Islamic Republic of Iran 2013a, 2013b).
183. These banks are *Bank Maskan* (housing), *Bank Keshavarzi* (agriculture), *Bank Sanat va Ma'dan* (industry and mining), and Export Development Bank. Available data are reported. Data are from *Bank Markazi Jomhouri-ye Islami-ye Iran* (Central Bank of the Islamic Republic of Iran) and available from ibid.
184. Keshavarzian 2007, 107–108.
185. The number of wholesale units grew 494 percent in urban areas and 273 percent in rural areas over this decade. Time-series and subnational data on individuals and establishments engaged in wholesale and retail trade are available from *Markaz-e Amar-e Iran* (Government of the Islamic Republic of Iran 2013b).
186. Evans 1979, 108.
187. All variables are measured per capita and logged to reduce skew.
188. Emrani 2013; Mohammadi 2013; Moqimi 2008.
189. Rahnema 1995, 141. See also Taghiyareh and Hekmat 2004; Amid and Hadjikhani 2005; Behkish 2005; Karimi et al. 2010; Momeni 2012.
190. Karshenas and Pesaran 1995, 97.
191. Ibid.
192. Hakimian 2007, 13.
193. Russia Today 2015.

CHAPTER 5

1. However, see Lahiri-Dutt 2014. On the environmental effects of mining see Bhushan and Hazra 2008. On mining and conflict see Kapur, Gawande, and Satyanath 2012; Hoelscher, Miklian, and Vadlamannati 2012.
2. The bulk of political economy research either focuses on all income generated from natural resources or one resource in particular, most often oil. Few studies offer detailed comparisons of specific types of natural resources. Two recent exceptions are Mitchell 2011 and Saylor 2014.

3. World Bank 2007, 61. There is large variation in how researchers choose to define "dependence" on natural resources. Many adopt a threshold of at least 10 percent of GDP.
4. See Saez 2000.
5. See Kale and Mazaheri 2014.
6. Recent work examines the effects of natural resources on subnational units. See Caselli and Michaels 2009; Goldberg, Wibbels, and Mvukiyehe 2008; Walker 2013.
7. On the controlled comparison methodology see George and Bennett 2004, ch. 8. The method of controlled comparison is particularly useful when comparing subnational units, given the ability to hold constant many country-level variables (see Snyder 2001). In the version that I employ, the researcher identifies two cases with divergent outcomes that are similar in all but one factor, the proposed causal variable (a resource windfall). The comparison of the two cases allows the researcher to test for the "treatment" effect of the proposed causal variable while "controlling" for other factors that could affect the outcome. The material used to construct a controlled comparison comes from secondary sources, economic data, newspapers, media reports, and interviews conducted in Bihar, Jharkhand, and New Delhi. The sample of subjects I identified was a group of individuals who possess in-depth knowledge and experience with government policy making, mining, commercial law, or entrepreneurial activities. These individuals belonged to the following occupational strata: officials with the state or central government, mining industry lobbyists, employees of large and small private mining firms, corporate attorneys, accountants, business association representatives, SME owners and workers in nonmining sectors, development consultants and researchers, and India-based academics, journalists, and activists. Snowball sampling led to the identification of additional interview subjects. A few respondents who discussed the situation in Jharkhand requested anonymity. See the appendix for a list of interviews cited in this chapter.
8. According to the 2011 Indian Census, Bihar has a population of 104 million and Jharkhand has a population of 33 million.
9. In 2000, Madhya Pradesh was bifurcated to create Chhattisgarh. In 2000, Uttar Pradesh was bifurcated to create Uttarakhand. See Tillin 2013 on these states. In 2014, Andhra Pradesh was bifurcated to create Telangana.
10. Mukherji and Mukherji 2012 and Santra, Kumar, and Bagaria 2014 focus on the effects of bifurcation on Bihar, whereas I look at both states.
11. See Prakash 2001; Shah 2010; Tillin 2013; Amit Prakash, personal interview.
12. Prasad 1997; Sinha 1981.
13. Yadava 1981, 49–53.
14. Prasad 1997, 23.
15. Yadava 1981, 48.
16. Shah 2010, 6.
17. Prakash 2001, 27–28.
18. Sushil Kumar, personal interview.
19. Mohammed Khan, personal interview. See also interview with Barkunwar Gargoi (MLA from Majhgaon, Jharkhand) in *Hindustan*. November 9, 2000: 10.
20. Ibid.
21. Ibid.
22. R. U. Singh, personal interview. Nitish Kumar, chief minister of Bihar as of writing, confessed: "I was initially opposed to the bifurcation of Bihar not only due

to long association of the people of the two states, but also for economic losses to Bihar after carving out of Jharkhand" (Outlook 2012).

23. *Hindustan*. November 16, 2000: 4.
24. Awadhesh Singh, personal interview; Sweta Ray, personal interview.
25. Harishwar Dayal, personal interview.
26. Data are from IndiaStat 2015, "Selected State-wise Amount of Royalty Paid for Mining of Coal in India" (various years).
27. The Pioneer 2012b. In the South Asian numbering system, one "crore" is equal to ten million and one "lakh" is equal to one hundred thousand. "Rs" is shorthand for Indian Rupees.
28. The Chiria mines located in East Singhbhum district have the second-largest deposits of high-grade iron ore in the world.
29. See Bhushan and Hazra 2008, 159–162.
30. Mukherji and Mukherji 2012, 12.
31. Data are from IndiaStat 2015, "Status of Registered Micro, Small and Medium Enterprises in Jharkhand (2006–2007)" and "Status of Registered Micro, Small and Medium Enterprises in Bihar (2006–2007)." About 96 percent of Bihar's MSMEs are categorized as "micro," whereas 80 percent of Jharkhand's MSMEs fall into this category. Data are from IndiaStat 2015, "State-wise Number of Registered Micro, Small and Medium Enterprises (MSMEs) in India (2008–2009 to 2011–2012)."
32. World Bank 2009.
33. If we compare these values to the country-level *start-up regulation* measure used in Chapter 2, Bihar and Jharkhand would rank in the bottom 20 percent of the world in 2012.
34. Indrajit Sinha, personal interview.
35. Rahul Singh, personal interview.
36. Lalit Kumar, personal interview.
37. Dilip Kumar, personal interview.
38. Deen Bandhu Singh, personal interview.
39. Arun Agarwal, personal interview.
40. Rahul Singh, personal interview.
41. Ibid.
42. Himanshu Adukia, personal interview.
43. Rohit Pandey, personal interview.
44. M. L. Siddiqui, personal interview.
45. Dilip Kumar, personal interview.
46. Sengupta 2014.
47. The "registering property" measure assumes a firm with fifty employees. The "paying taxes" measure assumes a firm with sixty employees. The "trading across borders" measure assumes a firm with sixty employees. The "closing a business" measure assumes a firm with 201 employees (refer to World Bank 2009). However, about 80 percent of all industrial units in Jharkhand are MSMEs (data from Jharkhand Small Industries Association) and there is an average of just 4.8 employees at each MSME and across all sectors (author's calculation using data in Khare 2013).
48. "SSI" is an older term used by the Indian government. It is today's equivalent of an MSME.

49. Data are from IndiaStat 2015, "Statewise Loans and Advances of Public Sector Banks to Small Scale Industries in Eastern India (2001 to 2009), Amount Outstanding at the End of March"; author's calculations.
50. Government of India 1976.
51. Bhatt and Thorat 2001, 74–75.
52. Data are from IndiaStat 2015, "State/Population Group-wise Number of Reporting Offices, Aggregate Deposits and Gross Bank Credit of Regional Rural Banks in Eastern India" (December 2001 to March 2012); author's calculations.
53. Ibid.
54. Ibid.
55. World Bank 2007, 73–74.
56. Ibid., 50 and 73–74.
57. Government of Jharkhand 2012, 61.
58. Rahul Ahuja, personal interview.
59. Rahul Singh, personal interview.
60. Ram Gulab, personal interview.
61. Harishwar Dayal, personal interview.
62. The Telegraph 2005.
63. Arun Agarwal, personal interview.
64. Mayank Shekhar, personal interview.
65. Lalit Kumar, personal interview.
66. Satyajit Kumar Singh, personal interview.
67. Amit Prakash, personal interview; K. K. Srivastava, personal interview.
68. Arun Kumar Khemka, personal interview.
69. World Bank 2007, 73–74.
70. Thakur 2002.
71. Chakravorti 2005.
72. Government of Jharkhand 2012, 81–82.
73. Ojha 2012.
74. The Pioneer 2013.
75. Ojha 2012; italics added.
76. Rahul Singh, personal interview.
77. Rahul Ahuja, personal interview.
78. Sanjay Kumar, personal interview.
79. Rahul Singh, personal interview.
80. Arun Kumar Khemka, personal interview.
81. Arun Agarwal, personal interview.
82. Ashim Mitra, personal interview.
83. Raj 2011.
84. Dilip Kumar, personal interview.
85. Ibid.; Satyajit Kumar Singh, personal interview.
86. Sengupta 2013, 2014.
87. Sengupta 2015.
88. It is only in recent years that Iran's private sector is allowed to play a small role in the country's oil industry, and these activities are closely regulated by the government.
89. Bhushan and Hazra 2008, 164–165.
90. World Bank 2007, 65.
91. Himanshu Adukia, personal interview.

92. Bhushan and Hazra 2008, 162.
93. Ibid.
94. Lahiri-Dutt, Krishnan, and Ahmad 2012, 40.
95. World Bank 2007, 116–117.
96. Bajaj and Yardley 2012.
97. Shrivastava 2012.
98. Bajaj and Yardley 2012.
99. UPI 2013.
100. Sharma 2012.
101. Ibid.
102. Singh 2013.
103. Anand 2012.
104. Sharma 2012. The Central Government had previously considered deallocating a block given to Abhijeet Group in 2005 because progress was not made (Arya 2012).
105. Sharma 2012.
106. The Times of India 2013.
107. The Economic Times-The Times of India 2013.
108. Samik Chowdhury, personal interview; Chandra Bhushan, personal interview.
109. Amit Prakash, personal interview.
110. The Economic Times-The Times of India 2012.
111. India Today 2010.
112. The Times of India 2012.
113. Koda reportedly did this through the activities of a businessman and confidante, Vijay Joshi. According to *The Times of India*, immediately after Vini Iron was sold to Vijay Joshi, the firm "won the state's backing while also bagging the Rajhara North block" (Chauhan 2012).
114. The owner of Rungta Mines stated that he had met Koda three times before, and "that purpose of meeting was to request him (to) expedite allotment of Param Baljori iron ore mine through cabinet as it was pending before the Jharkhand government since long, that after his meeting with Madhu Koda, it was cleared by the cabinet" (Kirpal 2012a).
115. Kirpal 2012c.
116. The Pioneer 2012a.
117. Bajaj and Yardley 2012. Beyond launching an official investigation and moving forward with deallocations, the Central Government's response to Coalgate was to set up auctions for coal blocks. However, they were only open to government-run firms. In November 2012, the Ministry of Coal banned government-owned firms from establishing joint ventures with private firms (UPI 2013).
118. Ramesh Sharan, personal interview.
119. DNA India 2012.
120. Kirpal 2012b.
121. Bajaj and Yardley 2012.
122. Three of these changes were to President's Rule, whereby the Central Government takes over direct control of the state.
123. With the exception of Madhu Koda, who was a political Independent, the BJP or JMM has ruled Jharkhand since independence. Arjun Munda and Shibu Soren each have served three separate tenures as chief ministers, and Shibu Soren's son Hemant served one term.

124. Ross 2001b, 40–41.
125. Shaibal Gupta, personal interview.
126. Nirmal Barman, personal interview.
127. Harishwar Dayal, personal interview.
128. R. D. Singh, personal interview.
129. Rahul Singh, personal interview; Arun Kumar Khemka, personal interview; R. D. Singh, personal interview.
130. Satyajit Kumar Singh, personal interview.
131. Arun Agarwal, personal interview.
132. Dinu, personal interview.
133. Nandy, personal interview.
134. S. Sharma, personal interview.
135. Baiju, personal interview.
136. Dinesh Kumar Verma, personal interview.
137. M. B. Chand, personal interview; Bijay Srivastava, personal interview; Maganlal Gupta, personal interview.
138. Dinu, personal interview.
139. Jashnuppal, personal interview.
140. Rupesh Punjabi, personal interview.
141. S. Sinha, personal interview.
142. Jashnuppal, personal interview.
143. Dinu, personal interview.
144. Maganlal Gupta, personal interview; Arvind Kumar, personal interview; Ravi, personal interview.
145. S. Sharma, personal interview.
146. Dinu, personal interview.
147. Rupesh Punjabi, personal interview.
148. Vikas Sharma, personal interview.
149. Baiju, personal interview; Ashok Kumar, personal interview.
150. Ranjan and Vasudev Prasad, personal interview.
151. Mahendar Sharma, personal interview; Ajay Kumar Sinha, personal interview.
152. Ibid.
153. Mahendar Sharma, personal interview; also Arvind Kumar, personal interview.
154. Awadhesh Singh, personal interview; Baiju, personal interview.
155. Rahul Singh, personal interview.
156. Kanhaiya Singh, personal interview.
157. NDTV 2010.
158. Bhushan and Hazra 2008, 181.
159. Singh 2011.
160. Kumar 2011.
161. DNA India 2011.
162. Bhushan and Hazra 2008, 17–18.
163. Singh 2011.
164. The Telegraph 2012. Santosh Tiwary, Vice-President of the East Central Railways Employee Union, said, "A racket involving the RPF [Railway Protection Force] and the GRP [Government Railway Police] is into the coal theft. Coal companies and railways are also involved in this racket." See Singh 2011.
165. Bharti 1989, 2353.

166. Ramesh Sharan, personal interview.
167. Jain 2006.
168. Sushil Kumar, personal interview.
169. According to the Ministry of MSMEs, an SSI seeks registration "to avail some benefits, incentives or support given either by the Central or State Government." The benefits offered by the Central Government are mainly related to credit, interest rates, and exemption from excise or other taxes. State governments "have their own package of facilities and incentives for small scale. They relate to development of industrial estates, tax subsidies, power tariff subsidies, capital investment subsidies and other support." See http://dcmsme.gov.in/howto-setup/grgxx01x.htm.
170. Data are from IndiaStat 2015, "District-wise Number of SSI Registered in Bihar" and "Statewise Total Number of Small Scale Industrial Units Functioning and Number of Units Registered in India" (various years); author's calculations.
171. Prebifurcation data were compiled using substate (district-level) data and grouping each district into the future state to which it would eventually belong.
172. Data are from IndiaStat 2015, "Statewise Distribution of Registration of New Non Government Companies in Eastern India (2000–2001 to 2010–2011)."
173. Data are from IndiaStat 2015, "State-wise Percentage Growth Rate of Per Capita Net State Domestic Product (NSDP) in India {(At Constant 2004–05 Prices) (2000–2001 to 2013–2014—as on 01.03.2014)}."
174. World Bank 2007, 63. To date, there is no study that seeks to systematically link Jharkhand's postbifurcation developmental pathway to its mineral wealth. However, some have speculated about a link (see Corbridge 2008; Nathan and Dayal 2009).
175. See Singh and Stern 2013; Mazaheri et al. 2013; D. M. Diwakar, personal interview.
176. Chandra Bhushan, personal interview; Satyaki Roy, personal interview; Pyare Lal, personal interview.
177. Mohammed Khan, personal interview.

CHAPTER 6

1. Portions of this chapter appeared in Mazaheri 2013. I thank the publishers of *Business and Politics* for allowing me to present the material here.
2. World Bank 2015.
3. Throughout the chapter, the term *monarchy* refers to dynastic monarchies that exercise absolute rule. For a discussion about the categorization of monarchies see Lucas 2004.
4. Herb 1999; Brownlee 2007. See work by Gelbach and Keefer 2011, which shows how an "institutionalized ruling party," the Chinese Communist Party, is able to commit to private investors.
5. Tullock 1987.
6. Anderson 1991, 4; italics added.
7. Bjornskov and Kurrild-Klitgaard 2008.
8. The persistence of economic familism in Saudi Arabia echoes research about the ways in which interpersonal relationships shape the construction of economic institutions (Granovetter 1992). See also Greif 1994, who argues that collective cultural beliefs can induce cooperation among individuals.
9. Benedict 1968; Wong 1993; Salstrom 1994.
10. James 2008.

11. Fukuyama 1995, 95. Researchers continue to investigate whether economic familism produces positive or negative outcomes. Most research centers on the role of family ownership in the private sector. See the review and analysis in Bertrand and Schoar 2006.
12. Niblock and Malik 2007, 30 and 106.
13. Ibid., 173; Looney 2004; Askari 2009, 412.
14. Luciani 2005, 172.
15. Niblock and Malik 2007, 187.
16. Malik 2004, 136.
17. Hertog 2011, 163–164.
18. World Bank 2011.
19. World Bank 2006.
20. World Bank 2015.
21. World Economic Forum 2013.
22. Al-Nefaee 2005, 79–80.
23. Interview with member of business association, Dubai, United Arab Emirates, June 2, 2009.
24. Clary and Karlin 2011, 18.
25. The Banker 2007.
26. Asiamoney 2008.
27. Hertog 2011, 255.
28. Ibid.
29. Al-Akoum 2009; Rivlin 2009, 235–236; Hertog 2011, 254.
30. Al-Salloum 1999; Alsaleh 2006.
31. Data to produce Figure 6.1 are from Kingdom of Saudi Arabia 2010.
32. SAGIA defines small-sized enterprises as those with fewer than sixty employees and medium-sized enterprises as those with fewer than one hundred employees.
33. Author's calculation using data from Kingdom of Saudi Arabia 2010.
34. Author's calculation using data from Al-Harthi 2000, 124.
35. Kingdom of Saudi Arabia 2010.
36. There is variation over time among nonoil sectors. For instance, more small firms have entered the textile sector and, to a lesser extent, the wood products sector, but this is not the case in the paper and leather products sectors.
37. Almarshad 2011, 331–332.
38. Almutery 2009, 17–41.
39. Hertog 2011, ch. 5.
40. Al-Nefaee 2005, 84.
41. Niblock and Malik 2007, 168–169.
42. Saudi Economic Survey 2011a.
43. Ramady 2005, 196–197; Saudi Economic Survey 2011b. In 2008, SMEs comprised 94.2 percent of all private-sector enterprises and provided 62.5 percent of all job opportunities. See Saudi Economic Survey 2011c.
44. Author's calculation using data from Kingdom of Saudi Arabia 2006, 2007, 2008, 2009.
45. Allam 2013.
46. Ibid.
47. AME Info 2010.
48. Chaudhry 1997; Krimly 1993; Niblock and Malik 2007.
49. Hertog 2007, 540, 545, and 547.
50. See Field 1984.

51. Krimly 1993, 376–381.
52. Ibid., 357–358.
53. Parasie and Nair 2010.
54. Al-Awaji 1971; cited in Wilson 2004, 13. In Saudi Arabia, "The family is the primary basis of identity and status for the individual and the immediate focus of individual loyalty" (Stenslie 2012, 79).
55. The Economist 2015.
56. Sfakianakis 2009.
57. Arab News 2013. See also Davis, Pitts, and Cormier 2000.
58. Daghsh 2004; cited in Ramady 2005, 200. See also Malik 2004, 134.
59. Hertog 2007, 555 and 540.
60. Ramady 2005, 201.
61. Saudi Gazette 2012.
62. Ramady 2009.
63. Ibid.
64. Washington Post nd.
65. Ramady 2005, 201.
66. Stenslie 2012, 39 and 57.
67. Al-Rasheed 2002, 220; see also Sharaf 2001.
68. Al-Rasheed 2002, 154.
69. Hertog 2007, 553.
70. Ibid., 545.
71. Rivlin 2009, 237.
72. Stenslie 2012, 59.
73. Wilson 2008, 134–135.
74. Wilson 2004, 26–27.
75. Hertog 2007, 555.
76. See Stenslie 2012, 59–60. Government publications try to suggest otherwise. See, for instance, Kingdom of Saudi Arabia 2007. This may be slowly changing. An economist stated, "Female entrepreneurship in Saudi Arabia is on the rise. Women have been successful in lobbying the government for greater freedoms to conduct business. Many have been successful in encouraging new policies, and providing greater opportunities for females to open businesses. This is a very positive sign" (Interview with economist, Dubai, United Arab Emirates, June 1, 2009). See also International Crisis Group 2004, 24–25.
77. Niblock and Malik 2007, 206.
78. Hertog 2011, 250–251 and 257–258. An economist stated, "The small business sector . . . does not yet possess a well-organized interest in Saudi Arabia. It is not like the West where you have a culture of organization among small businesses and annual meetings of owners and the publication of journals and all" (Interview with economist, Dubai, United Arab Emirates, May 28, 2009).
79. Samuel Huntington coined the phrase "king's dilemma" to describe the fundamental problem that monarchies face when considering modernization (1968, 177). See Anderson 1991 on Huntington's argument.
80. For a discussion of Middle Eastern monarchies and the prospects for political liberalization see Lucas 2004, 112–117 and Luciani 2007.
81. Olson 1993.
82. Haber, Razo, and Maurer 2003, 4.
83. North and Weingast 1989.
84. Landes and Posner 1975.

85. Williamson 1983; Haber, Razo, and Maurer 2003.

CHAPTER 7

1. Wilson and Graham 1994, 180.
2. Klapper and Parker 2011.
3. Ross 2008.
4. Simmons 2016.
5. Karl 2004; Caselli and Michaels 2009; Vicente 2010.
6. Fisman and Svensson 2007.
7. Şeker and Yang 2012.
8. Rose-Ackerman 1997, 56.
9. Morris and Klesner 2010.
10. Jamal and Nooruddin 2010.
11. Levi, Sacks, and Tyler 2009.
12. McNally 1997.
13. The research on natural resources and conflict does not make this assumption, but instead examines a wide range of qualities about resources (e.g., value per weight, point source vs. diffuse, lootable vs. nonlootable). See Auty 2004; Le Billon 2001; Ross 2004.
14. For more on Gulf states see Hvidt 2011.
15. This is similar to the logic in Dunning 2005, which argues that leaders of resource-dependent states are less likely to seek economic diversification if the political opposition is strong.
16. See Smith 2007; Morrison 2012; Basedau and Lay 2009.
17. See Helmke and Levitsky 2004, 2006.
18. Ross 2012, 215.
19. See Hvidt 2007 for more details about the Dubai model. See Gelb 2011 on economic diversification.

BIBLIOGRAPHY

Abrahamian, Ervand. 1978. "The Political Challenge." *MERIP Reports* 69: 3–8.

Abrahamian, Ervand. 1979. "The Causes of the Constitutional Revolution in Iran." *International Journal of Middle East Studies* 10 (August): 381–414.

Abrahamian, Ervand. 1982. *Iran between Two Revolutions*. Princeton, NJ: Princeton University Press.

Abrahamian, Ervand. 1993. *Khomeinism: Essays on the Islamic Republic*. Berkeley and Los Angeles: University of California Press.

Abrahamian, Ervand. 2013. *The Coup: 1953, the CIA, and the Roots of Modern U.S.-Iranian Relations*. New York: The New Press.

Acemoglu, Daron. 2008. "Oligarchic versus Democratic Societies." *Journal of the European Economic Association* 6 (March): 1–44.

Acemoglu, Daron, and James Robinson. 2012. *Why Nations Fail: The Origins of Power, Prosperity, and Poverty*. New York: Crown Publishers.

Acemoglu, Daron, Collier, Paul, Johnson, Simon, Klein, Michael, and Graeme Wheeler. 2013. "A Review of Doing Business." Open Letter. May 12.

Achen, Christopher. 2002. "Toward a New Political Methodology: Microfoundations and Art." *Annual Review of Political Science* 5: 423–450.

Acs, Zoltan, Carlsson, Bo, and Charlie Karlsson, eds. 2011. *Entrepreneurship, Small and Medium-Sized Enterprises and the Macroeconomy*. Cambridge: Cambridge University Press.

Aghion, Philippe, and Peter Howitt. 1998. *Endogenous Growth Theory*. Cambridge, MA: MIT Press.

Al-Akoum, Ibrahim. 2009. "Privatization in Saudi Arabia: Is Slow Beautiful?" *Thunderbird International Business Review* 51 (September–October): 427–440.

Al-Awaji, Ibrahim. 1971. "Bureaucracy and Society in Saudi Arabia." PhD diss., University of Virginia.

Alef. 2012a. "Taharokat-e Mashkuk Dar Bazaar-e Tehran [Suspicious Activities in the Tehran Bazaar]." October 3.

Alef. 2012b. "Dar Khiyabanhaye Atraf-e Bazar-e Tehran Che Gozasht [What Happened around the Streets of the Tehran Bazaar]." October 3.

Alef. 2012c. "Asgaroladi: Qovai Seh Ganeh Be Faryad-e E'teraz Bazarian Tavajoh Konand [Asgaroladi: Three Branches of Government Should Pay Attention to the Merchants' Protest]." October 4.

Alef. 2012d. "Tavafoq Etehadieha-ye Senfi Baraye Bazgasht-e Bazar Az Ruz-e Shanbe [Trade Unions Agreed to Reopen the Bazaar from Saturday]." October 4.

Alef. 2012e. "Tavanmandsazi-ye Bakhsh-e Khosusi Vazifeh-ye Siasatgozaran-ye Keshvar Ast [The Empowerment of the Private Sector Is the Duty of the Country's Policymakers]." December 31.

Alexeev, Michael, and Robert Conrad. 2009. "The Elusive Curse of Oil." *Review of Economics and Statistics* 91 (August): 586–598.
Al-Harthi, Mohammad. 2000. "The Political Economy of Labor in Saudi Arabia." PhD diss., Binghamton University.
Alizadeh, Parvin. 2003. "Iran's Quandary: Economic Reforms and the 'Structural Trap.'" *The Brown Journal of World Affairs* 9 (Winter–Spring): 267–281.
Allam, Abeer. 2013. "Arab Finance and Banking: Saudi Small Business Remains Reliant on Friends and Family for Funds." *Financial Times*, May 28.
Almarshad, Sultan. 2011. "The Impact of Good Governance and Decentralization Reforms on the Effectiveness of Local Authorities: The Case of Saudi Municipalities." PhD diss., University of Connecticut.
Almutery, Mohammed. 2009. "Locational Determinants of Foreign Direct Investment in the Kingdom of Saudi Arabia." PhD diss., Virginia Commonwealth University.
Al-Nefaee, Saad Mohammed. 2005. "Determinants of Disaggregate Private Investment in an Oil-Based Economy: The Case of Saudi Arabia." PhD diss., Colorado State University.
Al-Rasheed, Madawi. 2002. *A History of Saudi Arabia*. New York: Cambridge University Press.
Alsaleh, Shakir Bin Ahmed. 2006. "Diversification as the Nucleus of Economic Reform Policies in the Kingdom of Saudi Arabia: A Tripartite Study." PhD diss., Howard University.
Al-Salloum, Tariq. 1999. "Policy Choices in Developing Countries: The Case of Privatization in Saudi Arabia." PhD diss., George Mason University.
AME Info. 2010. "Middle East SMEs Report High Levels of Business Confidence." July 21.
Amid, Javad, and Amjad Hadjikhani. 2005. *Trade, Industrialization and the Firm in Iran: The Impact of Government Policy on Business*. London: I. B. Tauris.
Amin, Mohammad, and Simeon Djankov. 2009. "Natural Resources and Reforms." World Bank Policy Research Working Paper Number 4882, The World Bank, Washington, DC.
Amirahmadi, Hooshang. 1986. "Regional Planning in Iran: A Survey of Problems and Policies." *Journal of Developing Areas* 20 (July): 501–530.
Amjad, Mohammed. 1989. *Iran: From Royal Dictatorship to Theocracy*. New York: Greenwood Press.
Amsden, Alice, DiCaprio, Alisa, and James Robinson. 2012. *The Role of Elites in Economic Development*. Oxford: Oxford University Press.
Amuzegar, Jahangir. 1991. *The Dynamics of the Iranian Revolution: The Pahlavis' Triumph and Tragedy*. Albany: State University of New York Press.
Amuzegar, Jahangir. 1992. "The Iranian Economy before and after the Revolution." *Middle East Journal* 46 (Summer): 413–425.
Amuzegar, Jahangir. 1999. "Khatami and the Iranian Economy at Mid-term." *Middle East Journal* 53 (Autumn): 534–552.
Amuzegar, Jahangir. 2000. "Iran's 'Virtual Democracy' at a Turning Point." *SAIS Review* 20 (2): 93–109.
Anand, Utkarsh. 2012. "SC Questions Govt on Coal Blocks Allotment, Strongly Defends CAG." *Indian Express*, September 14.
Anderson, Lisa. 1991. "Absolutism and the Resilience of Monarchy in the Middle East." *Political Science Quarterly* 106 (Spring): 1–15.
Arab News. 2013. "Saudi Women's Role in Family Businesses Still Limited." April 10.

Arkhipov, Ilya. 2010. "Russian Revamp of Legal System May Free 100K Entrepreneurs." *Bloomberg News*, July 2.
Arrow, Kenneth, Cropper, Maureen, Eads, George, Hahn, Robert, Lave, Lester, Noll, Roger, Portney, Paul, Russell, Milton, Schmalensee, Richard, Smith, V. Kerry, and Robert Stavins. 1996. "Is There a Role for Benefit-Cost Analysis in Environmental, Health, and Safety Regulation?" *Science* 272 (April): 221.
Arya, Shishir. 2012. "MOEF Eases Norms for Coal Mine Expansion." *The Times of India*, December 27.
Asiamoney. 2008. "Saudi Unfastens Financial Sector." April 1.
Ashraf, Ahmad. 1995. "From the White Revolution to the Islamic Revolution." In *Iran After the Revolution: Crisis of an Islamic State*, ed. Saeed Rahnema and Sohrab Behdad, 22–44. London: I. B. Tauris.
Askari, Hossein. 2004. "Iran's Economic Policy Dilemma." *International Journal* 59 (Summer): 655–668.
Askari, Hossein. 2009. "Fixing the Mideast's Economies." *Current History* (December): 410–416.
Audretsch, David, Keilback, Max, and Erik Lehmann. 2006. *Entrepreneurship and Economic Growth*. New York: Oxford University Press.
Auty, Richard. 2001. "The Political Economy of Resource-Driven Growth." *European Economic Review* 45 (May): 839–846.
Auty, Richard. 2004. "Natural Resources and Civil Strife: A Two-Stage Process." *Geopolitics* 9 (1): 29–49.
Ayyagari, Meghana, Beck, Thorsten, and Asli Demirgüç-Kunt. 2005. *Small and Medium Enterprises across the Globe*. Washington, DC: The World Bank.
Ayyagari, Meghana, Demirgüç-Kunt, Asli, and Vojislav Maksimovic. 2011. "Small vs. Young Firms across the World: Contribution to Employment, Job Creation, and Growth." World Bank Policy Research Working Paper Number 5631, The World Bank, Washington, DC.
Bajaj, Vikas, and Jim Yardley. 2012. "Scandal Poses a Riddle: Will India Ever Be Able to Tackle Corruption?" *New York Times*, September 15.
Baland, Jean Marie, and Patrick Francois. 2000. "Rent-Seeking and Resource Booms." *Journal of Development Economics* 61 (April): 527–542.
Baldwin, George. 1967. *Planning and Development in Iran*. Baltimore: Johns Hopkins Press.
The Banker. 2007. "Economic Cities Point to a Sea Change in Thinking." November 1.
Banks, Arthur, and Kenneth Wilson. 2013. "Cross-National Time-Series Data Archive." Databanks International, Jerusalem.
Barma, Naazneen, Kaiser, Kai, Le, Tuan Minh, and Lorena Viñuela. 2012. *Rents to Riches? The Political Economy of Natural Resource-Led Development*. Washington, DC: The World Bank.
Barro, Robert. 2001. "Human Capital and Growth." *American Economic Review* 91 (May): 12–17.
Basedau, Matthias, and Jann Lay. 2009. "Resource Curse or Rentier Peace? The Ambiguous Effects of Oil Wealth and Oil Dependence on Violent Conflict." *Journal of Peace Research* 46 (November): 757–776.
Bashiriyeh, Hossein. 1984. *The State and Revolution in Iran, 1962–1982*. New York: St. Martin's.
Bates, Robert. 1981. *Markets and States in Tropical Africa*. Berkeley and Los Angeles: University of California Press.

Baumol, William. 2002. *The Free-Market Innovation Machine: Analyzing the Growth Miracle of Capitalism*. Princeton, NJ: Princeton University Press.

Bayat, Mangol. 1991. *Iran's First Revolution: Shi'ism and the Constitutional Revolution of 1905–1909*. New York: Oxford University Press.

Beblawi, Hazem, and Giacomo Luciani. 1987. *The Rentier State*. New York: Croom Helm.

Beck, Thorsten. 2007. "Financing Constraints of SMEs in Developing Countries: Evidence, Determinants, and Solutions." Working Paper, The World Bank, Washington, DC.

Beck, Thorsten, Demirgüç-Kunt, Asli, and Ross Levine. 2003. "Small and Medium Enterprises, Growth, and Poverty: Cross-Country Evidence." World Bank Policy Research Working Paper Number 3178, The World Bank, Washington, DC.

Behdad, Sohrab. 2000. "From Populism to Economic Liberalism: The Iranian Predicament." In *Iran's Economy: Dilemma of an Islamic State*, ed. Parvin Alizadeh, 100–144. London: I. B. Tauris.

Behdad, Sohrab. 2001. "Khatami and His 'Reformist' Economic (Non-)Agenda." *MERIP Reports*, May 21.

Behdad, Sohrab, and Farhad Nomani. 2009. "What a Revolution! Thirty Years of Social Class Reshuffling in Iran." *Comparative Studies of South Asia, Africa, and the Middle East* 29 (1): 84–104.

Beheshti, Mohammad. 2003. "Iran's Economic Development and Structural Change in Human Resources." In *Iran Encountering Globalization: Problems and Prospects*, ed. Ali Mohammadi, 149–179. New York: Routledge Curzon.

Behkish, M. M. 2005. *Eqtesad-e Iran* [Iran's Economy]. Tehran: Ney Publishing.

Bellin, Eva R. 2002. *Stalled Democracy: Capital, Labor, and the Paradox of State-Sponsored Development*. Ithaca, NY: Cornell University Press.

Benedict, Burton. 1968. "Family Firms and Economic Development." *Southwestern Journal of Anthropology* 24 (Spring): 1–19.

Bermeo, Sarah. 2016. "Aid Is Not Oil: Donor Utility, Heterogeneous Aid, and the Aid-Democratization Relationship." *International Organization* 70 (Winter): 1–32.

Bertrand, Marianne, and Antoinette Schoar. 2006. "The Role of Family in Family Firms." *Journal of Economic Perspectives* 20 (Spring): 73–96.

Bharti, Indu. 1989. "Usurpation of the State: Coal Mafia in Bihar." *Economic and Political Weekly* 24 (October): 2353.

Bhatt, Nitin, and Y. S. P. Thorat. 2001. "India's Regional Rural Banks: The Institutional Dimension of Reforms." *Journal of Microfinance* 3 (1): 65–94.

Bhushan, Chandra, and Monali Zeya Hazra. 2008. *Rich Lands Poor People: Is "Sustainable" Mining Possible?* New Delhi: Centre for Science and Environment.

Bjornskov, Christian, and Peter Kurrild-Klitgaard. 2008. "Economic Growth and Institutional Reform in Modern Monarchies and Republics: A Historical Cross-Country Perspective." Working Paper Number 08-15, Department of Economics, University of Aarhus.

Bjorvatn, Kjetil, and Kjetil Selvik. 2008. "Destructive Competition: Factionalism and Rent-Seeking in Iran." *World Development* 36 (November): 2314–2324.

Bjorvatn, Kjetil, Farzanegan, Mohammed Reza, and Friedrich Schneider. 2012. "Resource Curse and Power Balance: Evidence from Oil-Rich Countries." *World Development* 40 (July): 1308–1316.

Blomström, Magnus, and Ari Kokko. 2007. "From Natural Resources to High-Tech Production: The Evolution of Industrial Competitiveness in Sweden and Finland." In *Natural Resources: Neither Curse nor Destiny*, ed. Daniel Lederman and William Maloney, 213–256. Palo Alto, CA and Washington, DC: Stanford University Press and the World Bank.

Bodin, Jean. [1576] 1967. *Six Books of the Commonwealth*. Translated by M. J. Tooley. Oxford: Basil Blackwell.

Bonine, Michael. 1979. "The Morphogenesis of Iranian Cities." *Annals of the Association of American Geographers* 69 (2): 208–224.

Bottazzi, Laura, and Giovanni Perri. 2003. "Innovation and Spillovers in Regions: Evidence from European Patent Data." *European Economic Review* 47 (August): 687–710.

Bozorgmehr, Najmeh. 2009. "A Thirst to Reach the Top in Iran." *Financial Times*, May 6.

Braunerhjelm, Pontus, and Johan Eklund. 2013. "Taxes, Tax Administrative Burdens and New Firm Formation." CESIS Working Paper 312, Swedish Entrepreneurship Forum and Royal Institute of Technology.

Bravo-Ortega, Claudio, and Jose de Gregorio. 2007. "The Relative Richness of the Poor? Natural Resources, Human Capital, and Economic Growth." In *Natural Resources: Neither Curse nor Destiny*, ed. Daniel Lederman and William Maloney, 71–99. Palo Alto, CA and Washington, DC: Stanford University Press and the World Bank.

Breyer, Stephen. 1982. *Regulation and Its Reform*. Cambridge, MA: Harvard University Press.

Brownlee, Jason. 2007. "Hereditary Succession in Modern Autocracies." *World Politics* 59 (July): 595–628.

Bruhn, Miriam, and David McKenzie. 2013. "Entry Regulation and Formalization of Microenterprises in Developing Countries." World Bank Policy Research Working Paper Number 6507, The World Bank, Washington, DC.

Brunnschweiler, C. N., and E. H. Bulte. 2008. "The Resource Curse Revisited and Revised: A Tale of Paradoxes and Red Herrings." *Journal of Environmental Economics and Management* 55 (May): 248–264.

Bunce, Valerie. 2001. "Democratization and Economic Reform." *Annual Review of Political Science* 4: 43–65.

Calì, Massimiliano, and Kunal Sen. 2011. "Do Effective State Business Relations Matter for Economic Growth? Evidence from Indian States." *World Development* 39 (September): 1542–1557.

Cammett, Melani. 2007. *Globalization and Business Politics in Arab North Africa: A Comparative Perspective*. New York: Cambridge University Press.

Carey, Jane Perry Clark, and Andrew Carey. 1975. "Industrial Growth and Development Planning in Iran." *Middle East Journal* 29 (Winter): 1–15.

Caselli, Francesco, and Tom Cunningham. 2009. "Leader Behaviour and the Natural Resource Curse." *Oxford Economic Papers* 61: 628–650.

Caselli, Francesco, and Guy Michaels. 2009. "Do Oil Windfalls Improve Living Standards? Evidence from Brazil." NBER Working Paper Number 15550, National Bureau of Economic Research, Cambridge, MA.

Chakravorti, Tapan. 2005. "Jharkhand Sets Up Project Clearance Agency." *Business Standard*, October 26.

Chaudhry, Kiren Aziz. 1997. *The Price of Wealth: Economies and Institutions in the Middle East*. Ithaca, NY: Cornell University Press.

Chauhan, Neeraj. 2012. "Madhu Koda Next in Firing Line? CBI Finds Irregularities in Block Allotments during Ex-CM's Tenure?" *The Economic Times-Times of India*, September 9.

Chen, An. 2002. "Capitalist Development, Entrepreneurial Class, and Democratization in China." *Political Science Quarterly* 117 (Autumn): 401–422.

Cipolla, Carlo. 1969. *Literacy and Development in the West*. Harmondsworth, UK: Penguin.

Clary, Christopher, and Mara Karlin. 2011. "Saudi Arabia's Reform Gamble." *Survival* 53 (October–November): 15–20.
Clawson, Patrick. 1988. "Islamic Iran's Economic Politics and Prospects." *Middle East Journal* 42 (Summer): 371–388.
Collier, David. 2011. "Understanding Process Tracing." *PS: Political Science and Politics* 44 (October): 823–830.
Collier, Paul. 2006. "Is Aid Oil? An Analysis of Whether Africa Can Absorb More Aid." *World Development* 34 (September): 1482–1497.
Collier, Paul. 2010. *The Plundered Planet*. New York: Oxford University Press.
Corbridge, Stuart. 2008. "Jharkhand: A Series of Unfortunate Events." India in Transition, Center for the Advanced Study of India at the University of Pennsylvania, January 24. https://casi.sas.upenn.edu/iit/scorbridge.
Corden, W. Max, and J. Peter Neary. 1982. "Booming Sector and De-industrialization in a Small Open Economy." *Economic Journal* 92 (December): 825–848.
Crystal, Jill. 1995. *Oil and Politics in the Gulf: Rulers and Merchants in Kuwait and Qatar*. New York: Cambridge University Press.
Dabashi, Hamid. 2007. *Iran: A People Interrupted*. New York: The New Press.
Daghsh, Muna. 2004. "Family Companies Have SR250 Billion in Investments." *Arab News*, April 2.
Dasgupta, Kunal. 2010. "The Missing Middle in Developing Countries Revisited." Working Paper, University of Toronto.
Davis, John, Pitts, Elye, and Keely Cormier. 2000. "Challenges Facing Family Companies in the Gulf Region." *Family Business Review* 13 (September): 217–238.
Davis, Kevin, and Michael Kruse. 2007. "Taking the Measure of Law: The Case of the Doing Business Project." *Law and Social Inquiry* 32 (Fall): 1095–1119.
Deacon, Robert, and Ashwin Rode. 2012. "Rent Seeking and the Resource Curse." Manuscript, University of California, Santa Barbara. http://www.econ.ucsb.edu/~deacon/RentSeekingResourceCurse%20Sept%2026.pdf
Denoeux, Guilain. 1993. *Urban Unrest in the Middle East: A Comparative Study of Informal Networks in Egypt, Iran, and Lebanon*. Albany: State University of New York Press.
De Soto, Hernando. 1989. *The Other Path*. New York: Basic Books.
Dias, Joilson, and John McDermott. 2006. "Institutions, Education, and Development: The Role of Entrepreneurs." *Journal of Development Economics* 80 (August): 299–328.
Diop, Ndiamé, Marotta, Daniela, and Jaime de Melo. 2012. *Natural Resource Abundance, Growth, and Diversification in the Middle East and North Africa: The Effects of Natural Resources and the Role of Policies*. Washington, DC: The World Bank.
Djankov, Simeon, McLiesh, Caralee, and Andrei Shleifer. 2007. "Private Credit in 129 Countries." *Journal of Financial Economics* 84 (2): 299–329.
Djankov, Simeon, La Porta, Rafael, Lopez-de-Silanes, Florenzio, and Andrei Shleifer. 2002. "The Regulation of Entry." *Quarterly Journal of Economics* 116 (February): 1–37.
Djankov, Simeon, Ganser, Tim, McLiesh, Caralee, Ramalho, Rita, and Andrei Shleifer. 2010. "The Effect of Corporate Taxes on Investment and Entrepreneurship." *American Economic Journal: Macroeconomics* 2 (July): 31–64.
DNA India. 2011. "Coal Worth Millions Plundered in Eastern and Central India." September 19.
DNA India. 2012. "Mining Companies Funding Politicians." September 11.
Doner, Richard, and Ben Ross Schneider. 2000. "Business Associations and Economic Development: Why Some Associations Contribute More Than Others." *Business and Politics* 2 (3): 2000.

Dorraj, Manouchehr. 1990. *From Zarathustra to Khomeini: Populism and Dissent in Iran*. Boulder, CO: Lynne Rienner.
Duflo, Esther, Banerjee, Abhijit, Glennerster, Rachel, and Cynthia G. Kinnan. 2013. "The Miracle of Microfinance? Evidence from a Randomized Evaluation." NBER Working Paper Number 18950, National Bureau of Economic Research, Cambridge, MA.
Dunning, Thad. 2005. "Resource Dependence, Economic Performance, and Political Stability." *Journal of Conflict Resolution* 49 (August): 451–482.
Dunning, Thad. 2008. *Crude Democracy: Natural Resource Wealth and Political Regimes*. New York: Cambridge University Press.
The Economic Times-The Times of India. 2012. "Madhu Koda Gets Bail, Will Remain in Prison in Connection with Other Cases." April 6.
The Economic Times-The Times of India. 2013. "Coalgate Scam: CBI Files FIR Accusing Naveen Jindal and Dasari Narayan Rao of Fraud." June 12.
The Economist. 2015. "The Algosaibi Affair: A Saudi Saga." March 15.
Eeles, Miranda. 2004. "Iran's Entrepreneurs See Calm Future." *BBC News*, May 24.
Ehsani, Kaveh. 1994. "'Tilt but Don't Spill': Iran's Development and Reconstruction Dilemma." *Middle East Report* 191: 16–21.
Ehteshami, Anoushiravan, and Mahjoob Zweiri. 2007. *Iran and the Rise of Its Neo-conservatives: The Politics of Tehran's Silent Revolution*. New York: I. B. Tauris.
Emrani, Abolhasan. 2013. "Agha-ye Rais Jomhur Motevali Bakhsh-e Khosusi dar Keshvar Kist? [Mr. President Who Is in Charge of the Country's Private Sector?]." *Alef*, January 9.
Etemaad. 2009. "Moshkel-e Sherkat-haye Shabiheh Dolati [The Problem with Quasi-Governmental Firms]." July 4.
Etemaad. 2013. "'Omdeh Ma'vqat-e Banki Marbut be Bakhsh-e Khosusi va Omumi Ast [The Majority of Late Payments to the Banks Are Pertaining to the Private and Public Sectors]." January 31.
Evans, Peter. 1979. *Dependent Development: The Alliance of Multinational, State, and Local Capital in Brazil*. Princeton, NJ: Princeton University Press.
Evans, Peter. 1995. *Embedded Autonomy: States and Industrial Transformation*. Princeton, NJ: Princeton University Press.
Evans, Peter, and John D. Stephens. 1988. "Studying Development since the Sixties: The Emergence of a New Comparative Political Economy." *Theory and Society* 17 (September): 713–745.
Falkinger, Josef, and Volker Grossmann. 2005. "Distribution of Natural Resources, Entrepreneurship, and Economic Development: Growth Dynamics with Two Elites." IZA Discussion Papers Number 1756, Institute for the Study of Labor, Bonn.
Fardmanesh, Mohsen. 1991. "Dutch Disease Economics and the Oil Syndrome: An Empirical Study." *World Development* 19 (June): 711–717.
Farhi, Farideh. 2013. "Is Iran on the Brink?" *Lobe Log/IPS News*, January 16.
Farzin, Hossein. 1995. "The Political Economy of Foreign Exchange Control." In *Iran After the Revolution: Crisis of an Islamic State*, ed. Saeed Rahnema and Sohrab Behdad, 174–200. London: I. B. Tauris.
Field, Michael. 1984. *The Merchants: The Big Business Families of Saudi Arabia and the Gulf States*. Woodstock, NY: Overlook Press.
Fisman, Raymond, and Jakob Svensson. 2007. "Are Corruption and Taxation Really Harmful to Growth? Firm Level Evidence." *Journal of Development Economics* 83 (May): 63–75.

Foran, John. 1993. *Fragile Resistance: Social Transformation in Iran from 1500 to the Revolution*. Boulder, CO: Westview Press.

Foran, John. 1994. *A Century of Revolution: Social Movements in Iran*. Minneapolis: University of Minnesota Press.

Forbes. 2007. "The Entrepreneur's Challenge in Nigeria." March 6.

Frankel, Jeffrey. 2010. "The Natural Resource Curse: A Survey." NBER Working Paper Number 15836, National Bureau of Economic Research, Cambridge, MA.

Friedman, Eric, Johnson, Simon, Kaufmann, Daniel, and Pablo Zoido-Lobaton. 2000. "Dodging the Grabbing Hand: The Determinants of Unofficial Activity in 69 Countries." *Journal of Public Economics* 76 (June): 459–493.

Frye, Timothy. 2002. "Capture or Exchange? Business Lobbying in Russia." *Europe-Asia Studies* 54 (November): 1017–1036.

Fukuyama, Francis. 1995. "Social Capital and the Global Economy: A Redrawn Map of the World." *Foreign Affairs* 74 (September–October): 89–103.

Galindo, Arturo José, and Alejandro Micco. 2007. "Creditor Protection and Credit Response to Shocks." *World Bank Economic Review* 21 (September): 413–438.

Ganji, Manouchehr. 2002. *Defying the Iranian Revolution: From a Minister to the Shah to a Leader of Resistance*. Westport, CT: Greenwood.

Gelb, Alan. 2011. "Economic Diversification in Resource Rich Countries." Working Paper, Center for Global Development, Washington, DC.

Gelb, Alan, and Associates. 1988. *Oil Windfalls: Blessing or Curse?* New York: Oxford University Press and the World Bank.

Gelb, Alan, and Sina Grasmann. 2010. "How Should Oil Exporters Spend Their Rents?" Working Paper Number 221, Center for Global Development, Washington, DC.

Gelbach, Scott, and Philip Keefer. 2011. "Investment without Democracy: Ruling-Party Institutionalization and Credible Commitment in Autocracies." *Journal of Comparative Economics* 39 (June): 123–139.

Gelman, Andrew. 2007. "Clustered Standard Errors vs. Multilevel Modeling." Blog entry. http://andrewgelman.com/2007/11/28/clustered_stand/.

George, Alexander, and Andrew Bennett. 2004. *Case Studies and Theory Development in the Social Sciences*. Cambridge, MA: MIT Press.

George, Alexander, and Timothy McKeown. 1985. "Case Studies and Theories of Organizational Decision Making." In *Advances in Information Processing in Organizations (Vol. 2)*, ed. Robert Coulam and Richard Smith, 21–58. Greenwich, CT: JAI Press.

Ghandchi-Tehrani, Davoud. 1982. "Bazaaris and Clergy: Socio-economic Origins of Radicalism and Revolution in Iran." PhD diss., City University of New York.

Ghasimi, M. R. 1992. "The Iranian Economy after the Revolution: An Economic Appraisal of the Five-Year Plan." *International Journal of Middle East Studies* 24 (November): 599–614.

Ghods, M. Reza. 1989. *Iran in the Twentieth Century: A Political History*. Boulder, CO: Lynne Rienner.

Ghorayshi, Parvin. 1981. "Capitalism in Rural Iran." *MERIP Reports* 98 (July–August): 28–30.

Gladstone, Rick. 2012. "Merchants Reopen in Tehran, with Police Watching for More Protests." *New York Times*, October 4.

Goldberg, Ellis, Wibbels, Erik, and Eric Mvukiyehe. 2008. "Lessons from Strange Cases: Democracy, Development, and the Resource Curse in the U.S. States." *Comparative Political Studies* 41 (April): 477–514.

Golkar, Saeid. 2012. "Paramilitarization of the Economy: The Case of Iran's Basij Militia." *Armed Forces and Society* 38 (April): 625–648.
Government of Australia. 2007. *Best Practice Regulation Handbook*. Canberra, Australia.
Government of India. 1976. "Chapter 1, Preliminary, The Regional Rural Banks Act, 1976 Act No. 21 of 1976 [February 9, 1976]." New Delhi, India.
Government of the Islamic Republic of Iran. 2013a. "Amar [Statistics]." Bank Markazi Jomhouri-ye Islami-ye Iran [Central Bank of the Islamic Republic of Iran]. Tehran, Iran.
Government of the Islamic Republic of Iran. 2013b. "Amar [Statistics]." Markaz-e Amar-e Iran [Statistical Centre of Iran]. Tehran, Iran. http://www.amar.sci.org.ir.
Government of Jharkhand. 2012. *Jharkhand Industrial Policy-2012*. Ranchi, India.
Graham, Robert. 1980. *Iran: The Illusion of Power*. New York: St. Martin's.
Granovetter, Mark. 1992. "Economic Institutions as Social Constructions." *Acta Sociologica* 35 (January): 3–11.
Greenwood, Scott. 2008. "Bad for Business? Entrepreneurs and Democracy in the Arab World." *Comparative Political Studies* 41 (June): 837–860.
Greif, Avner. 1994. "Cultural Beliefs and the Organization of Society." *Journal of Political Economy* 102 (October): 912–950.
Grindle, Merilee. 2005. "When Good Policies Go Bad, Then What?" Paper presented at the University of Manchester, School of Environment and Development, May.
Haber, Stephen, and Victor Menaldo. 2011a. "Do Natural Resources Fuel Authoritarianism? A Reappraisal of the Resource Curse." *American Political Science Review* 105 (February): 1–26.
Haber, Stephen, and Victor Menaldo. 2011b. "Haber and Menaldo 2011 APSR Datasets." Stanford University Social Science History Program Datasets.
Haber, Stephen, Razo, Armando, and Noel Maurer. 2003. *The Politics of Property Rights*. New York: Cambridge University Press.
Haddad, Bassam. 2012. *Business Networks in Syria: The Political Economy of Authoritarian Resilience*. Palo Alto, CA: Stanford University Press.
Haggard, Stephan, and Robert Kauffman. 1992. *The Politics of Economic Adjustment*. Princeton, NJ: Princeton University Press.
Haggard, Stephan, MacIntyre, Andrew, and Lydia Tiede. 2008. "The Rule of Law and Economic Development." *Annual Review of Political Science* 11: 205–234.
Hakimian, Hassan. 2007. "Institutional Change, Policy Challenges, and Macroeconomic Performance: Case Study of the Islamic Republic of Iran (1979–2004)." Commission on Growth and Development Working Paper Number 26, The World Bank, Washington, DC.
Halliday, Fred. 1979. *Iran: Dictatorship and Development*. New York: Penguin Press.
Hallward-Driemeier, Mary, and Lant Pritchett. 2011. "How Business Is Done and the Doing Business Indicators: The Investment Climate When Firms Have Climate Control." World Bank Policy Research Working Paper Number 5563, The World Bank, Washington, DC.
Hamshahri. 2009. "Negarani-ye Fa'alan-e Eqtesadi, Delgarmi Vazir [Concern of Economic Actors, an Optimistic Minister]." March 8.
Hamshahri. 2011. "Ahmadinejad: Asnaf az Ashoftegi-ye Bazar Khoshhal Nemishavand [Ahmadinejad: Guilds Are Not Happy with Chaos in the Market]." June 22.
Hamshahri. 2012. "Darkhast-e Bazar va Fa'alan-e Eqtesadi Az Dolat Baraye Raf'e Moshkelat [The Bazaar and Economic Actors Demand that the Government Eliminate Problems]." October 5.

Harding, Torfinn, and Anthony Venables. 2013. "The Implications of Natural Resource Exports for Non-Resource Trade." OxCarre Research Paper Number 103, Oxford Centre for the Analysis of Resource Rich Economies, University of Oxford.

Harris, Kevan. 2013a. "A Fistful of Tomans." *London Review of Books* 35 (January 24).

Harris, Kevan. 2013b. "The Rise of the Subcontractor State: Politics of Pseudo-privatization in the Islamic Republic of Iran." *International Journal of Middle East Studies* 45 (February): 45–70.

Hausmann, Ricardo, and Roberto Rigobon. 2003. "An Alternative Interpretation of the 'Resource Curse': Theory and Policy Implications." NBER Working Paper Number 9424, National Bureau of Economic Research, Cambridge, MA.

Heller, Michael, and Rebecca Eisenberg. 1998. "Can Patents Deter Innovation? The Anticommons in Biomedical Research." *Science* 280 (May): 698–701.

Hellman, Joel, and Daniel Kaufmann. 2003. "The Inequality of Influence." Paper presented at Stanford Corruption Workshop, January 30–31.

Helmke, Gretchen, and Steven Levitsky. 2004. "Informal Institutions and Comparative Politics: A Research Agenda." *Perspectives on Politics* 2 (December): 725–740.

Helmke, Gretchen, and Steven Levitsky, eds. 2006. *Informal Institutions and Democracy: Lessons from Latin America*. Baltimore: Johns Hopkins University Press.

Herb, Michael. 1999. *All in the Family: Absolutism, Revolution, and Democracy in the Middle Eastern Monarchies*. Albany: SUNY Press.

Hertog, Steffen. 2007. "Shaping the Saudi State: Human Agency's Shifting Role in Rentier-state Formation." *International Journal of Middle East Studies* 39 (4): 539–563.

Hertog, Steffen. 2011. *Princes, Brokers, and Bureaucrats: Oil and the State in Saudi Arabia*. Ithaca, NY: Cornell University Press.

Heuty, Antoine. 2012. "A Ticking Bomb? Iran's Oil and Gas Management." *Revenue Watch Institute Briefing*, February.

Hirschman, Albert O. 1958. *The Strategy of Economic Development*. New Haven, CT: Yale University Press.

Hoelscher, Kristian, Miklian, Jason, and Krishna Chaitanya Vadlamannati. 2012. "Hearts and Mines: A District-Level Analysis of the Maoist Conflict in India." *International Area Studies Review* 15 (June): 141–160.

Hsiao, Hsin-Huan Michael, and Hagen Koo. 1997. "The Middle Classes and Democratization." In *Consolidating the Third Wave Democracies: Themes and Perspectives*, ed. Larry Diamond, Marc F. Plattner, Yun-han Chu, and Hung-mao Tien, 312–334. Baltimore: Johns Hopkins University Press.

Hsieh, Chang-Tai, and Benjamin Olken. 2014. "The Missing 'Missing Middle.'" *Journal of Economic Perspectives* 28 (Summer): 89–108.

Humphreys, Macartan, Sachs, Jeffrey, and Joseph Stiglitz. 2007. *Escaping the Resource Curse*. New York: Columbia University Press.

Hunter, Shireen T. 1992. *Iran after Khomeini*. Westport, CT: Praeger.

Huntington, Samuel. 1952. "The Marasmus of the ICC: The Commission, the Railroads, and the Public Interest." *Yale Law Journal* 61 (April): 467–509.

Huntington, Samuel. 1968. *Political Order in Changing Societies*. New Haven, CT: Yale University Press.

Hvidt, Martin. 2007. "The Dubai Model: An Outline of Key Components of the Development Process in Dubai." Centre for Contemporary Middle East Studies Working Paper Number 12, University of Southern Denmark.

Hvidt, Martin. 2011. "Economic and Institutional Reforms in the Arab Gulf Countries." *Middle East Journal* 65 (Winter): 85–102.

IndiaStat. 2015. "Various Statistics." Datanet India. http://www.indiastat.com.

India Today. 2010. "Radia Brokered Mining Deal with Koda: CBI Report." November 23.
International Crisis Group. 2004. "Can Saudi Arabia Reform Itself?" *International Crisis Group Middle East Report*, 28.
International Finance Corporation. 2010. *Scaling-Up SME Access to Financial Services in the Developing World*. Washington, DC: IFC-World Bank Group.
International Labour Organization. 2002. *Women and Men in the Informal Economy: A Statistical Picture*. Geneva: International Labour Organization.
Issawi, Charles. 1978. "The Iranian Economy 1925–1975: Fifty Years of Economic Development." In *Iran under the Pahlavis*, ed. George Lenczowski, 129–166. Stanford, CA: Hoover Institution Press.
Jahan-Parvar, Mohammad R., and Hassan Mohammadi. 2009. "Oil Prices and Competitiveness: Time Series Evidence from Six Oil-Producing Countries." *Journal of Economic Studies* 36 (1): 98–118.
Jahan-Parvar, Mohammad R., and Hassan Mohammadi. 2011. "Oil Prices and Real Exchange Rates in Oil-Exporting Countries: A Bounds Testing Approach." *Journal of Developing Areas* 45 (Fall): 309–318.
Jain, Sonu. 2006. "Officer Exposes Mining Corruption, Coal Minister Soren Wants Him Out." *The Indian Express*, August 20.
Jalali-Naini, Ahmad. 2008. "Capital Accumulation, Financial Market Reform and Growth in Iran: Past Experience and Future Prospects." In *Iran and the 21*st *Century: Politics, Economics, and Conflict*, ed. Homa Katouzian and Hossein Shahidi, 217–242. New York: Routledge Press.
Jamal, Amaney, and Irfan Nooruddin. 2010. "The Democratic Utility of Trust: A Cross-National Analysis." *Journal of Politics* 72 (January): 45–59.
James, Harold. 2008. "Family Values or Crony Capitalism?" *Capitalism and Society* 3 (1): 1–28.
Jenanseft, Mohammad Sadiq. 2013. "Emruz va Fardai Qodrat-e Khosusi [Private Sector Power Today and Tomorrow]." *Donya-e Eqtesad*, January 20.
Jones Luong, Pauline, and Erika Weinthal. 2010. *Oil Is Not a Curse: Ownership Structure and Institutions in Soviet Successor States*. New York: Cambridge University Press.
Kale, Sunila, and Nimah Mazaheri. 2014. "Natural Resources, Development Strategies, and Lower Caste Empowerment in India's Mineral Belt: Bihar and Odisha during the 1990s." *Studies in Comparative International Development* 49 (July): 343–369.
Kamalian, Aminreza, Rashki, Maryam, and Mah Lagha Arbabi. 2011. "Barriers to Innovation among Iranian SMEs." *Asian Journal of Business Management* 3 (2): 79–90.
Kamrava, Mehran. 2004. "The Semi-Formal Sector and the Turkish Political Economy." *British Journal of Middle Eastern Studies* 31 (May): 63–87.
Kapur, Devesh, Gawande, Kishore, and Shanker Satyanath. 2012. "Renewable Resource Shocks and Conflict in India's Maoist Belt." CASI Working Paper Series Number 12-02, University of Pennsylvania.
Karimi, Saeid, Chizari, Mohammed, Biemans, Harm, and Martin Mulder. 2010. "Entrepreneurship Education in Iranian Higher Education: The Current State and Challenges." *European Journal of Scientific Research* 48 (1): 35–50.
Karl, Terry. 1997. *The Paradox of Plenty: Oil Booms and Petro-states*. Berkeley and Los Angeles: University of California Press.
Karl, Terry. 2004. "Oil-Led Development: Social, Political, and Economic Consequences." *Encyclopedia of Energy* 4: 661–672.
Karshenas, Massoud. 1990. *Oil, State, and Industrialization in Iran*. Cambridge: Cambridge University Press.

Karshenas, Massoud, and Hassan Hakimian. 2008. "Managing Oil Resources and Economic Diversification in Iran." In *Iran and the 21st Century: Politics, Economics, and Conflict*, ed. Homa Katouzian and Hossein Shahidi, 194–216. New York: Routledge Press.

Karshenas, Massoud, and M. Hashem Pesaran. 1995. "Economic Reform and the Reconstruction of the Iranian Economy." *Middle East Journal* 49 (Winter): 89–111.

Kayhan. 2012. "Ashofteh Bazar-e Arz: Che Dastha-ye Dar Kar Ast? [Currency Market Chaos: Whose Responsibility Is It?]." October 1.

Katouzian, Homa. 1981. *The Political Economy of Modern Iran: Despotism and Pseudo-modernism, 1926–1979*. New York: New York University Press.

Katouzian, Homa. 2009. *The Persians: Ancient, Mediaeval and Modern Iran*. New Haven, CT: Yale University Press.

Keddie, Nikki. 1980. *Iran: Religion, Politics, and Society*. London: Frank Cass.

Keddie, Nikki. 1983. "Iranian Revolutions in Comparative Perspective." *The American Historical Review* 88 (June): 579–598.

Keddie, Nikki, and Mehrdad Amanat. 1991. "Iran under the late Qajars, 1848–1922." In *The Cambridge History of Iran: From Nadir Shah to the Islamic Republic (Volume 7)*, ed. Peter Avery, Gavin Hambly, and Charles Melville, 174–212. New York: Cambridge University Press.

Kerr, William, Lincoln, William, and Prachi Mishra. 2011. "The Dynamics of Firm Lobbying." NBER Working Paper Number 17577, National Bureau of Economic Research, Cambridge, MA.

Keshavarzian, Arang. 2007. *Bazaar and State in Iran: The Politics of the Tehran Marketplace*. New York: Cambridge University Press.

Khajehpour, Bijan. 2007. "Iran: An Analysis of Economic Reforms and Developments." Manuscript.

Khare, Divy. 2013. "Jharkhand Focus on to Get 10 Lakhs Small Industries by 2022." *The Times of India*, March 25.

Kian-Thiebaut, Azadeh. 2001. "Mavan'e Karafarini dar Iran: Eqtesad-e Dolati va Na Amni-ye Hoquqi [Barriers to Entrepreneurship in Iran: The State Economy and Legal Insecurity]." *Iran Nameh* 19 (4).

Kingdom of Saudi Arabia. 2006. *Saudi Industrial Development Fund Annual Report, 1426/1427H (2006G)*. Riyadh, Saudi Arabia.

Kingdom of Saudi Arabia. 2007. *Saudi Industrial Development Fund Annual Report, 1427/1428H (2007G)*. Riyadh, Saudi Arabia.

Kingdom of Saudi Arabia. 2008. *Saudi Industrial Development Fund Annual Report, 1428/1429H (2008G)*. Riyadh, Saudi Arabia.

Kingdom of Saudi Arabia. 2009. *Saudi Industrial Development Fund Annual Report, 1430/1431H (2009G)*. Riyadh, Saudi Arabia.

Kingdom of Saudi Arabia. 2010. *Saudi Arabian Monetary Agency Annual Report*. Riyadh, Saudi Arabia.

Kirpal, Raman. 2012a. "Madhu Koda Connection Emerges in Coalgate Scam." *FirstPost*, September 6.

Kirpal, Raman. 2012b. "The Who's Who of Mining Are Funding India's Political Parties." *FirstPost*, September 11.

Kirpal, Raman. 2012c. "A Tata Coalgate? 999-yr Mine Lease at 25p a Bigha!" *FirstPost*, October 25.

Klapper, Leora, and Inessa Love. 2010. "The Impact of the Financial Crisis on New Firm Registration." World Bank Policy Research Working Paper Number 5444, The World Bank, Washington, DC.

Klapper, Leora, and Simon Parker. 2011. "Gender and the Business Environment for New Firm Creation." *World Bank Research Observer* 26 (August): 237–257.

Klapper, Leora, Laeven, Luc, and Raghuram Rajan. 2006. "Entry Regulation as a Barrier to Entrepreneurship." *Journal of Financial Economics* 82 (December): 591–625.

Kolstad, Ivar, and Arne Wiig. 2009. "Is Transparency the Key to Reducing Corruption in Resource-Rich Countries?" *World Development* 37 (March): 521–532.

Krimly, Rayed Khalid. 1993. "The Political Economy of Rentier States: A Case Study of Saudi Arabia in the Oil Era, 1950–1990." PhD diss., George Washington University.

Krueger, Anne. 1974. "The Political Economy of the Rent-Seeking Society." *American Economic Review* 64 (June): 291–303.

Krueger, Anne. 2014. "The Missing Middle." In *Economic Reform in India: Challenges, Prospects, and Lessons*, ed. Nicholas Hope, Anjini Kochar, Roger Noll, and T. N. Srinivasan, 299–318. New York: Cambridge University Press.

Kumar, Abhay. 2011. "Dhanbad's Dirty 'Black Diamond.'" *Deccan Herald*, February 5.

Lahiri-Dutt, Kuntala, ed. 2014. *The Coal Nation: Histories, Ecologies and Politics of Coal in India*. Aldershot, UK: Ashgate.

Lahiri-Dutt, Kuntala, Krishnan, Radhika, and Nesar Ahmad. 2012. "Land Acquisition and Dispossession: Private Coal Companies in Jharkhand." *Economic and Political Weekly* 47 (February): 39–45.

Landes, William, and Richard Posner. 1975. "The Independent Judiciary in an Interest-Group Perspective." NBER Working Paper Number 0110, National Bureau of Economic Research, Cambridge, MA.

La Porta, Rafael, and Andrei Shleifer. 2014. "Informality and Development." *Journal of Economic Perspectives* 28 (Summer): 109–126.

Le Billon, Philippe. 2001. "The Political Ecology of War: Natural Resources and Armed Conflicts." *Political Geography* 20 (June): 561–581.

Lederman, Daniel, and William Maloney, eds. 2007. *Natural Resources: Neither Curse nor Destiny*. Palo Alto, CA and Washington, DC: Stanford University Press and the World Bank.

Levi, Margaret, Sacks, Audrey, and Tom Tyler. 2009. "Conceptualizing Legitimacy, Measuring Legitimating Beliefs." *American Behavioral Scientist* 53 (November): 354–375.

Levine, Ross. 2005. "Does Firm Size Matter for Growth and Poverty Alleviation?" Paper presented at Brookings Blum Roundtable: The Private Sector in the Fight against Global Poverty, August 4.

Liu, Michael Tien-Lung. 1988. "States and Urban Revolutions: Explaining the Revolutionary Outcomes in Iran and Poland." *Theory and Society* 17 (March): 179–209.

Livermore, Michael, and Richard Revesz. 2013. *The Globalization of Cost-Benefit Analysis in Environmental Policy*. New York: Oxford University Press.

Lloyd Evans, Sally. 2008. "Geographies of the Contemporary Informal Sector in the Global South: Gender, Employment Relationships and Social Protection." *Geography Compass* 2 (November): 1885–1906.

Looney, Robert. 2004. "Saudization and Sound Economic Reforms: Are the Two Compatible?" *Strategic Insights* 3 (February).

Lowi, Miriam. 2009. *Oil Wealth and the Poverty of Politics: Algeria Compared*. New York: Cambridge University Press.

Lucas, Russell. 2004. "Monarchical Authoritarianism: Survival and Political Liberalization in a Middle Eastern Regime Type." *International Journal of Middle East Studies* 36 (February): 103–119.

Luciani, Giacomo. 2005. "Saudi Arabian Business: From Private Sector to National Bourgeoisie." In *Saudi Arabia in the Balance*, ed. Paul Aarts and Gerd Nonnemann, 144–181. New York: New York University Press.

Luciani, Giacomo. 2007. "Linking Economic and Political Reform in the Middle East: The Role of the Bourgeoisie." In *Debating Arab Authoritarianism: Dynamics and Durability in Nondemocratic Regimes*, ed. Oliver Schlumberger, 161–176. Stanford, CA: Stanford University Press.

Luebbert, Gregory. 1991. *Liberalism, Fascism, or Social Democracy: Social Classes and the Political Origins of Regimes in Interwar Europe*. New York: Oxford University Press.

Macher, Jeffrey, Mayo, John, and Mirjam Schiffer. 2011. "The Influence of Firms on Government." *The B.E. Journal of Economic Analysis and Policy* 11 (1): 1–25.

MacLachlan, K. S. 1991. "Economic Development 1921–1979." In *The Cambridge History of Iran: From Nadir Shah to the Islamic Republic (Volume 7)*, ed. Peter Avery, Gavin Hambly, and Charles Melville, 608–638. New York: Cambridge University Press.

Mahdavy, Hossein. 1970. "The Patterns and Problems of Economic Development in Rentier States: The Case of Iran." In *Studies in the Economic History of the Middle East: From the Rise of Islam to the Present Day*, ed. M. A. Cook, 428–467. London: Oxford University Press.

Maleki, Abbas. 2009. "Oil Economies and Social Welfare: Iran." Manuscript. Centro Studi di Politica Internazionale, Rome, Italy.

Malik, Monica. 2004. "The Role of the Private Sector." In *Economic Development in Saudi Arabia*, ed. Rodney Wilson, Abdullah Al-Salamah, Monica Malik, and Ahmed Al-Rajhi, 126–138. New York: Routledge Curzon.

Maloney, William. 2007. "Missed Opportunities: Innovation and Resource-Based Growth in Latin America." In *Natural Resources: Neither Curse nor Destiny*, ed. Daniel Lederman and William Maloney, 111–167. Palo Alto, CA and Washington, DC: Stanford University Press and The World Bank.

Marshall, Monty, and Keith Jaggers. 2012. "Polity IV Project: Political Regime Characteristics and Transitions, 1800–2011." Center for Systemic Peace.

Maxfield, Sylvia, and Ben Ross Schneider. 1997. *Business and the State in Developing Countries*. Ithaca, NY: Cornell University Press.

Mazaheri, Nimah. 2006. "State Repression in the Iranian Bazaar: The Anti-Profiteering Campaign and an Impending Revolution." *Iranian Studies* 39 (September): 401–414.

Mazaheri, Nimah. 2013. "The Saudi Monarchy and Economic Familism in an Era of Business Environment Reforms." *Business and Politics* 15 (3): 295–321.

Mazaheri, Nimah, Al-Dahdah, Edouard, Poundrik, Sandeep, and Soujanya Chodavarapu. 2013. "Leadership and Institutional Change in the Public Provision of Transportation Infrastructure: An Analysis of India's Bihar." *Journal of Development Studies* 49 (January): 19–35.

McNally, Kevin. 1997. *Corporate Venture Capital: Bridging the Equity Gap in the Small Business Sector*. London: Routledge.

Mehlum, Halvor, Moene, Karl, and Ragnar Torvik. 2006. "Institutions and the Resource Curse." *Economic Journal* 116 (January): 1–20.

Mehner, Harald. 1978. "Development and Planning in Iran after World War II." In *Iran under the Pahlavis*, ed. George Lenczowski, 167–199. Stanford, CA: Hoover Institution Press.

Menaldo, Victor. 2016. *The Institutions Curse: Natural Resources, Politics, and Development*. New York: Cambridge University Press.

Miller, Sam. 2013. "The Resource Curse: L.A. Teams Spend Big Money on Vets, but Young Talent Wins in Today's Game." *ESPN: The Magazine*, May 28.

Mir-Motahari, Seyed Ahmad. 2013. "Az Naft dar Sofre-ha ta Eqtesad-e Bedun-e Naft [From Oil on Tables to an Economy without Oil]." *Donya-e Eqtesad*, January 24.

Mitchell, Tim. 2011. *Carbon Democracy: Political Power in the Age of Oil*. New York: Verso.

Moaddel, Mansoor. 1986. "Shi'i Ulama and the State in Iran." *Theory and Society* 15 (July): 519–556.

Mohammadi, Ali Asghar. 2013. "Qavanin-e Dolati Mohemtarin Man'e Karafarini [Government Regulation Most Significant Barrier to Entrepreneurship]." *Hamshahri*, January 25.

Momeni, Farshad. 2012. "Tamam-e Siasathaye Eqtesadi ra Baznegari Konid [Review All Economic Policies]." *Etemaad*, March 14.

Moore, Barrington. 1966. *Social Origins of Dictatorship and Democracy*. Boston: Beacon Press.

Moore, Pete W. 2001. "What Makes Successful Business Lobbies? Business Associations and the Rentier State in Jordan and Kuwait." *Comparative Politics* 33 (January): 127–147.

Moore, Pete W. 2002. "Rentier Fiscal Crisis and Regime Stability: Business-State Relations in the Gulf." *Studies in Comparative International Development* 37 (Spring): 34–56.

Moore, Pete W. 2004. *Doing Business in the Middle East: Politics and Economic Crisis in Jordan and Kuwait*. New York: Cambridge University Press.

Moore, Pete W., and Bassel Salloukh. 2007. "Struggles under Authoritarianism: Regimes, States, and Professional Associations in the Arab World." *International Journal of Middle East Studies* 39: 53–76.

Moqimi, Seyyed Mohammed. 2008. "Qavanin-e Dolati; Man'e Karafarini Ast [Government Regulation is Barrier to Entrepreneurship]." *Hamshahri*, July 13.

Morris, Stephen, and Joseph Klesner. 2010. "Corruption and Trust: Theoretical Considerations and Evidence from Mexico." *Comparative Political Studies* 43 (10): 1258–1285.

Morrison, Kevin. 2009. "Oil, Nontax Revenue, and the Redistributional Foundations of Regime Stability." *International Organization* 63 (January): 107–138.

Morrison, Kevin. 2012. "Oil, Conflict, and Stability." Working Paper, University of Pittsburgh.

Mourougane, Annabelle. 2012. "Promoting SME Development in Indonesia." OECD Economics Department Working Papers Number 995, OECD Publishing.

Mozaffari, Mehdi. 1991. "Why the Bazar Rebels." *Journal of Peace Research* 28 (4): 377–391.

Mukherji, Arnab, and Anjan Mukherji. 2012. "Bihar: What Went Wrong? And What Changed?" Working Paper Number 2012-107, National Institute of Public Finance and Policy, New Delhi.

Naghshineh-Pour, Amir. 2008. "A Review and Analysis of Iran's Current Economic Status." MPRA Paper Number 13313, University of Munich.

Narayan, Deepa. 2002. *Empowerment and Poverty Reduction: A Sourcebook*. Washington, DC: The World Bank.

Nasr, Vali. 2000. "Politics within the Late-Pahlavi State: The Ministry of Economy and Industrial Policy, 1963–69." *International Journal of Middle East Studies* 32 (1): 97–122.

Nasr, Vali. 2009. *Forces of Fortune: The Rise of the New Muslim Middle Class and What It Will Mean for Our World*. New York: Free Press.

Nathan, Dev, and Harishwar Dayal. 2009. "Resource Curse and Jharkhand." *Economic and Political Weekly* 44 (December): 16–17.
NDTV. 2010. "Naxals Turn Mining Mafia in Jharkhand." April 20.
Niblock, Tim with Monica Malik. 2007. *The Political Economy of Saudi Arabia*. London: Routledge.
Nielsen, Lynge. 2011. "Classification of Countries Based on Their Level of Development: How It Is Done and How It Could Be Done." IMF Working Paper Number 11/31, The International Monetary Fund, Washington, DC.
Norris, P. 1999. *Critical Citizens: Global Support for Democratic Governance*. New York: Oxford University Press.
North, Douglass, and Barry Weingast. 1989. "Constitutions and Commitment: The Evolution of Institutions Governing Public Choice in Seventeenth Century England." *Journal of Economic History* 40 (December): 803–832.
Nosrati, Mehdi. 2013. "Khodahafezi ba Rant-e Nafti [Goodbye to Oil Rents]." *Donya-e Eqtesad*, January 22.
Ojha, Sanjay. 2012. "Single Window System a Distant Dream for Jharkhand Inc." *The Times of India*, December 13.
Okruhlik, Gwenn. 1999. "Rentier Wealth, Unruly Law, and the Rise of Opposition: The Political Economy of Oil States." *Comparative Politics* 31 (April): 295–315.
Olson, Mancur. 1965. *The Logic of Collective Action*. Cambridge, MA: Harvard University Press.
Olson, Mancur. 1982. *The Rise and Decline of Nations*. New Haven, CT: Yale University Press.
Olson, Mancur. 1993. "Dictatorship, Democracy and Development." *American Political Science Review* 87 (September): 567–576.
Outlook. 2012. "I Had Initially Opposed Bifurcation of Bihar: Nitish." May 7.
Pahlavi, Mohammed Reza. 1980. *Answer to History*. New York: Stein and Day.
Paine, Chris, and Erica Schoenberger. 1975. "Iranian Nationalism and the Great Powers: 1872–1954." *MERIP Reports* 37 (May): 3–28.
Papyrakis, Elissaios, and Reyer Gerlagh. 2005. "Natural Resources, Innovation and Growth." Paper presented at DEGIT Conference, June. http://degit.sam.sdu.dk/papers/degit_10/C010_054.pdf
Parasie, Nicolas, and Dinesh Nair. 2010. "Gulf Reforms Key to Private Equity Regaining Allure." *Reuters*, August 12.
Parsa, Misagh. 1988. "Theories of Collective Action and the Iranian Revolution." *Sociological Forum* 3 (Winter): 44–71.
Parsa, Misagh. 1989. *Social Origins of the Iranian Revolution*. New Brunswick, NJ: Rutgers University Press.
Parsa, Misagh. 1994. "Mosque of Last Resort: State Reform and Social Conflict in the Early 1960s." In *A Century of Revolution: Social Movements in Iran*, ed. John Foran, 135–159. Minneapolis: University of Minnesota Press.
Parsons, Talcott. 1951. *The Social System*. Glencoe, IL: Free Press.
Pearce, David. 1997. "Benefit-Cost Analysis, Environment, and Health in the Developed and Developing World." *Environment and Development Economics* 2 (May): 210–214.
Peregudov, Sergei, and Irina Semenenko. 1996. "Lobbying Business Interests in Russia." *Democratization* 3 (Summer): 115–139.
Pesaran, Evaleila. 2011. *Iran's Struggle for Economic Independence: Reform and Counter-Reform in the Post-Revolutionary Era*. New York: Routledge.
The Pioneer. 2012a. "Centre Against Jharkhand Land Allocation to Tata Steel." July 12.

The Pioneer. 2012b. "Crores Paid to Jharkhand as Royalty on Coal in Last Five Years." August 13.

The Pioneer. 2013. "CII Jharkhand Seeks Single Window Clearance System." April 12.

Ploeg, Frederick van der. 2011. "Natural Resources: Curse or Blessing?" *Journal of Economic Literature* 49 (June): 366–420.

Poteete, Amy R. 2009. "Is Development Path Dependent or Political? A Reinterpretation of Mineral-Dependent Development in Botswana." *Journal of Development Studies* 45 (April): 544–571.

Prakash, Amit. 2001. *Jharkhand: Politics of Development and Identity*. New Delhi: Orient Longman.

Prasad, K. N. 1997. *Bihar Economy through the Plans*. New Delhi: Northern Book Centre.

Press TV. 2009. "Iran Business Leaders Criticize Privatization Protocol." September 3.

PricewaterhouseCoopers. 2014. *Paying Taxes 2014: The Global Picture*. London: PricewaterhouseCoopers.

Rahnema, Saeed. 1995. "Continuity and Change in Industrial Policy." In *Iran after the Revolution: Crisis of an Islamic State*, ed. Saeed Rahnema and Sohrab Behdad, 129–149. London: I. B. Tauris.

Raj, Anand. 2011. "Power Drive for Industry Roadmap." *The Telegraph*, June 8.

Ramady, Mohamed. 2005. *The Saudi Arabian Economy: Policies, Achievements, and Challenge*. New York: Springer.

Ramady, Mohamed. 2009. "Saudi Family Businesses Sink under Debt." *The National*, July 4.

Ramazani, Rouhollah K. 1974. "Iran's 'White Revolution': A Study in Political Development." *International Journal of Middle East Studies* 5 (April): 124–139.

Resalat. 2013. "Bayad Eqtesad-e Moqavamati az Sho'ar be Sho'ur Beresad [The 'Resistance Economy' Should Be Understood Instead of Slogans]." February 7.

Rivlin, Paul. 2009. *Arab Economies in the Twenty-First Century*. New York: Cambridge University Press.

Robinson, James A., Torvik, Ragnar, and Thierry Verdier. 2006. "Political Foundations of the Resource Curse." *Journal of Development Economics* 79 (April): 447–468.

Rodrik, Dani. 2008. "Second-Best Institutions." NBER Working Paper Number 14050, National Bureau of Economic Research, Cambridge, MA.

Rose-Ackerman, Susan. 1997. "The Political Economy of Corruption." In *Corruption and the Global Economy*, ed. Kimberly Ann Elliot, 31–60. Washington, DC: Institute for International Economics.

Ross, Michael. 2001a. "Does Oil Hinder Democracy?" *World Politics* 53 (April): 325–361.

Ross, Michael. 2001b. *Timber Booms and Institutional Breakdown in Southeast Asia*. New York: Cambridge University Press.

Ross, Michael. 2004. "What Do We Know About Natural Resources and Civil War?" *Journal of Peace Research* 41 (May): 337–356.

Ross, Michael. 2008. "Oil, Islam, and Women." *American Political Science Review* 102 (February): 107–123.

Ross, Michael. 2009. "Oil and Democracy Revisited." Working Paper, Department of Political Science, UCLA.

Ross, Michael. 2012. *The Oil Curse: How Petroleum Wealth Shapes the Development of Nations*. Princeton, NJ: Princeton University Press.

Ross, Michael. 2013. "Oil and Gas Data, 1932–2011." Harvard Dataverse Network.

Ross, Michael, and Jørgen J. Andersen. 2014. "The Big Oil Change: A Closer Look at the Haber-Menaldo Analysis." *Comparative Political Studies* 47 (7): 993–1021.

Rotblat, Howard J. 1972. “Stability and Change in an Iranian Provincial Bazaar.” PhD diss., University of Chicago.

Rueschemeyer, Dietrich, Stephens, Evelyne Huber, and John Stephens. 1992. *Capitalist Development and Democracy.* Chicago: University of Chicago Press.

Russia Today. 2015. “Greater Freedom for Businesses—Best Response to Sanctions, Putin Says.” March 19.

Sachs, Jeffrey D. 1999. “Resource Endowments and the Real Exchange Rate: A Comparison of Latin America and East Asia.” In *Changes in Exchange Rates in Rapidly Developing Countries: Theory, Practice, and Policy Issues*, ed. Takatoshi Ito and Anne Krueger, 133–154. Chicago: University of Chicago Press.

Sachs, Jeffrey D. 2007. “How to Handle the Macroeconomics of Oil Wealth.” In *Escaping the Resource Curse*, ed. Macartan Humphreys, Jeffrey Sachs, and Joseph Stiglitz, 173–193. New York: Columbia University Press.

Saeidi, Ali. 2004. “The Accountability of Para-government Organizations (Bonyads): The Case of Iranian Foundations.” *Iranian Studies* 37 (Summer): 479–498.

Saez, Lawrence. 2002. *Federalism without a Centre: The Impact of Political and Economic Reform on India's Federal System.* New Delhi: Sage Publications.

Safari Esfahani, Bahareh. 2009. “Khosusihaye Natavan va Shabiheh Dolati-haye Porshetab [Weak Private Firms and Accelerating Quasi-Governmentals].” *Hamshahri*, October 6.

Sala-i-Martin, Xavier, and Arvind Subramanian. 2003. “Addressing the Natural Resource Curse: An Illustration from Nigeria.” NBER Working Paper Number 9804, National Bureau of Economic Research, Cambridge, MA.

Salehi Esfahani, Hadi. 2005. “Alternative Public Service Delivery Mechanisms in Iran.” *Quarterly Review of Economics and Finance* 45 (May): 497–525.

Salehi-Isfahani, Djavad. 1989. “The Political Economy of Credit Subsidy in Iran, 1973–1978.” *International Journal of Middle East Studies* 21 (August): 359–379.

Salstrom, Paul. 1994. *Appalachia's Path to Dependency: Rethinking a Region's Economic History, 1730–1940.* Lexington: University of Kentucky Press.

Santra, Swarup, Kumar, Rajesh, and Nidhi Bagaria. 2014. “Structural Change of Bihar Economy during 1999 to 2010: A District Level Analysis.” Social Science Research Network Working Paper Series.

Saudi Economic Survey. 2011a. “SMEs in Kingdom to Receive More Financial Support.” April 6.

Saudi Economic Survey. 2011b. “SMEs Key to Growth in Saudi Arabia.” April 20.

Saudi Economic Survey. 2011c. “Holding Back: State Spending Focus Restrains Private Sector Diversification.” June 1.

Saudi Gazette. 2012. “Syndicated Bank Lending in KSA Doubles in 2012.” November 12.

Saylor, Ryan. 2014. *State Building in Boom Times: Commodities and Coalitions in Latin America and Africa.* New York: Oxford University Press.

Schneider, Ben Ross. 1997. “Big Business and the Politics of Economic Reform: Confidence and Concertation in Brazil and Mexico.” In *Business and the State in Developing Countries*, ed. Sylvia Maxfield and Ben Ross Schneider, 191–215. Ithaca, NY: Cornell University Press.

Schneider, Friedrich, Buehn, Andreas, and Claudio Montenegro. 2010. “Shadow Economies All over the World.” World Bank Policy Research Working Paper Number 5356, The World Bank, Washington, DC.

Schumpeter, Joseph. [1911] 1934. *The Theory of Economic Development.* Cambridge, MA: Harvard University Press.

Scott, James. 1977. *The Moral Economy of the Peasant: Rebellion and Subsistence in Southeast Asia*. New Haven, CT: Yale University Press.
Şeker, Murat, and Judy Yang. 2012. "How Bribery Distorts Firm Growth: Differences by Firm Attributes." World Bank Policy Research Working Paper Number 6046, The World Bank, Washington, DC.
Selvik, Kjetil. 2005. "The Rise and Newfound Legitimacy of Iran's Industrial Bourgeoisie." Gulf Studies Number 2, University of Oslo.
Sengupta, Joy. 2013. "Single Window Back as Udyog Mitra." *The Telegraph*, April 13.
Sengupta, Joy. 2014. "Single Window Focus for Investors." *The Telegraph*, September 27.
Sengupta, Joy. 2015. "Focus on Single Window." *The Telegraph*, December 13.
Sfakianakis, John. 2009. "Challenges Facing Family Businesses in Saudi Arabia, Gulf." *Arab News*, June 18.
Shadlen, Kenneth. 2002. "Orphaned by Democracy: Small Industry in Contemporary Mexico." *Comparative Politics* 35 (October): 43–62.
Shah, Alpa. 2010. *In the Shadows of the State: Indigenous Politics, Environmentalism, and Insurgency in Jharkhand, India*. Durham, NC: Duke University Press.
Shambayati, Hootan. 1994. "The Rentier State, Interest Groups and the Paradox of Autonomy: State and Business in Turkey and Iran." *Comparative Politics* 26 (April): 307–331.
Shams Ardekani, Ali. 2013. "Budje va Bimari Holandi [The Budget and Dutch Disease]." *Donya-e Eqtesad*, January 7.
Sharaf, Sabri. 2001. *The House of Saud in Commerce: A Study of Royal Entrepreneurship in Saudi Arabia*. New Delhi: I. S. Publications.
Shargh. 2011. "Otagh Bayad Otagh Bashad Na Vezarat-e Eqtesad [The Chamber Should Be a Chamber, Not Ministry of the Economy]." June 14.
Sharma, Supriya. 2012. "Coalgate: One Family That Struck Gold in Coal Mines." *The Times of India*, September 5.
Shleifer, Andrei, and Robert Vishny. 1993. "Corruption." *Quarterly Journal of Economics* 108 (3): 599–617.
Shleifer, Andrei, and Robert Vishny. 1994. "Politicians and Firms." *Quarterly Journal of Economics* 109 (4): 995–1025.
Shrivastava, Suman. 2012. "People's Say in Coal Must." *The Telegraph*, September 12.
Simmons, Joel. 2016. "Resource Wealth and Women's Economic and Political Power in the U.S. States." *Comparative Political Studies* 49 (1): 115–152.
Singh, Gyanant. 2013. "Coalgate Scam: CBI Report Indicating Coal Block Allocations Done Arbitrarily Seem to Suggest the Probe Going the 2G Way." *India Today*, March 13.
Singh, N. K., and Nicholas Stern, eds. 2013. *The New Bihar: Rekindling Governance and Development*. Noida: HarperCollins India.
Singh, Rohit. 2011. "In Jharkhand, Coal Robbery Has Become a Way of Life." *India Today*, June 28.
Sinha, Amareshwar Prasad. 1981. "Economic Backwardness of Bihar and Certain Relevant Aspects of Census Data." In *Studies in Bihar's Economy and Society*, ed. M. N. Karna, 15–27. New Delhi: Concept.
Sinha, Aseema. 2005. "Understanding the Rise and Transformation of Business Collective Action in India." *Business and Politics* 7 (August): 1–35.
Slinko, Irina, Yakovlev, Evgeny, and Ekaterina Zhuravskaya. 2005. "Laws for Sale: Evidence from Russia." *American Law and Economics Review* 7 (Spring): 284–318.

Smith, Adam. [1776] 1976. *An Inquiry into the Nature and Causes of the Wealth of Nations*. Oxford: Clarendon Press.

Smith, Benjamin. 2004. "Oil Wealth and Regime Survival in the Developing World, 1960–1999." *American Journal of Political Science* 48 (2): 232–246.

Smith, Benjamin. 2007. *Hard Times in the Lands of Plenty: Oil Politics in Iran and Indonesia*. Ithaca, NY: Cornell University Press.

Snyder, Richard. 2001. "Scaling Down: The Subnational Comparative Method." *Studies in Comparative International Development* 36 (Spring): 93–110.

Spatafora, Nikola, and Andrew Warner. 1995. "Macroeconomic Effects of Terms-of-Trade Shocks: The Case of Oil-Exporting Countries." World Bank Policy Research Working Paper Number 1410, The World Bank, Washington, DC.

Stenslie, Stig. 2012. *Regime Stability in Saudi Arabia: The Challenge of Succession*. New York: Routledge.

Stigler, George. 1971. "The Theory of Economic Regulation." *The Bell Journal of Economics and Management Science* 2 (Spring): 3–21.

Stigler, George. 1974. "Free Riders and Collective Action: An Appendix to Theories of Regulation." *Bell Journal of Economics and Management Science* 5 (August): 359–365.

Stijns, Jean-Philippe, 2005. "Natural Resource Abundance and Economic Growth Revisited." *Resources Policy* 30 (June): 107–130.

Stone, Andrew, Levy, Brian, and Ricardo Paredes. 1992. "Public Institutions and Private Transactions: The Legal and Regulatory Environment for Business Transactions in Brazil and Chile." World Bank Policy Research Working Paper Number 891, The World Bank, Washington, DC.

Sunstein, Cass. 2003. *The Cost-Benefit State: The Future of Regulatory Protection*. Chicago: American Bar Association.

Szirmai, Adam, Naudé, Wim, and Micheline Goedhuys, eds. 2011. *Entrepreneurship, Innovation, and Economic Development*. New York: Oxford University Press.

Taghiyareh, Fataneh, and Sara Nouri Hekmat. 2004. "A New Entrepreneurial Model for Iranian Incubators." Manuscript, University of Tehran.

Tarrow, Sidney. 1995. "Bridging the Quantitative-Qualitative Divide in Political Science." *American Political Science Review* 89 (June): 471–474.

Thakur, Arun. 2002. "Jharkhand Industrial Policy Loses Its Meaning." *The Times of India*, April 2.

The Telegraph. 2005. "Sparks of Instinct in Samaritan Success." September 12.

The Telegraph. 2012. "Yashwant Wants CBI to Probe Coal Mafia—Plea to Tackle Dhanbad Racket." May 18.

The Times of India. 2012. "Coalgate: CBI Raids at Five Locations in Jharkhand." September 5.

The Times of India. 2013. "Coalgate: Naveen Jindal Group Misrepresented Facts, Says CBI." July 1.

Therborn, G. 1977. "The Rule of Capital and the Rise of Democracy." *New Left Review* 103 (May–June): 3–41.

Thorp, Rosemary, and Francisco Durand. 1997. "A Historical View of Business-State Relations: Colombia, Peru, and Venezuela Compared." In *Business and the State in Developing Countries*, ed. Sylvia Maxfield and Ben Ross Schneider, 216–236. Ithaca, NY: Cornell University Press.

Thurik, Roy, and Sander Wennekers. 2004. "Entrepreneurship, Small Business and Economic Growth." *Journal of Small Business and Enterprise Development* 11 (1): 140–149.

Tillin, Louise. 2013. *Remapping India: New States and their Political Origins*. Oxford: Oxford University Press.
Torvik, Ragnar. 2002. "Natural Resources, Rent Seeking and Welfare." *Journal of Development Economics* 67 (April): 455–470.
Treasury Board of Canada. 2007. *Canadian Cost-Benefit Analysis Guide: Regulatory Proposals*. Ottawa, Canada.
Tullock, Gordon. 1987. *Autocracy*. Dordrecht, The Netherlands: Kluwer Academic Publisher.
Tybout, James. 2000. "Manufacturing Firms in Developing Countries: How Well Do They Do, and Why?" *Journal of Economic Literature* 38 (March): 11–44.
Tybout, James. 2014. "The Missing Middle, Revisited." Working Paper, Pennsylvania State University.
UPI. 2013. "India to Auction Coal Blocks." January 3.
Van De Walle, Nicolas. 1994. "Political Liberation and Economic Reform in Africa." *World Development* 22 (April): 483–500.
Van Stel, André, Storey, David, and A. Roy Thurik. 2007. "The Effect of Business Regulations on Nascent and Young Business Entrepreneurship." *Small Business Economics* 28 (March): 171–186.
Vicente, Pedro. 2010. "Does Oil Corrupt? Evidence from a Natural Experiment in West Africa." *Journal of Development Economics* 92 (May): 28–38.
Walker, Annie. 2013. "An Empirical Analysis of Resource Curse Channels in the Appalachian Region." Working Paper, West Virginia University.
Washington Post. nd. "Saudi Arabia: Climate for Change."
Weinthal, Erika, and Pauline Jones Luong. 2006. "Combating the Resource Curse: An Alternative Solution to Managing Mineral Wealth." *Perspectives on Politics* 4 (March): 35–53.
Wennekers, Sander, and Roy Thurik. 1999. "Linking Entrepreneurship and Economic Growth." *Small Business Economics* 13: 27–55.
Weyland, Kurt. 2009. "The Rise of Latin America's Two Lefts: Insights from Rentier State Theory." *Comparative Politics* 41 (January): 145–164.
Weymouth, Stephen. 2012. "Firm Lobbying and Influence in Developing Countries: A Multilevel Approach." *Business and Politics* 14 (January): 1–26.
Wibbels, Erik, and Ellis Goldberg. 2007. "Natural Resources, Development and Democracy: The Quest for Mechanisms." Working Paper, Duke University and University of Washington.
Wilber, Donald N. 1948. *Iran: Past and Present*. Princeton, NJ: Princeton University Press.
Wildavsky, Aaron. 1966. "The Political Economy of Efficiency: Cost-Benefit Analysis, Systems Analysis, and Program Budgeting." *Public Administration Review* 26 (December): 292–310.
Williamson, Oliver. 1983. "Credible Commitments: Using Hostages to Support Exchange." *American Economic Review* 73 (September): 519–540.
Wilson, Peter, and Douglas Graham. 1994. *Saudi Arabia: The Coming Storm*. Armonk, NY: M. E. Sharpe.
Wilson, Rodney. 2004. *Economic Development in Saudi Arabia*. London: Routledge Curzon.
Wilson, Rodney. 2008. "Economic Governance and Reform in Saudi Arabia." In *Reform in the Middle East Oil Monarchies*, ed. Anoushiravan Ehteshami and Steven Wright, 132–152. Reading, UK: Ithaca Press.
Wong, Siu-Iun. 1993. "The Chinese Family Firm: A Model." *Family Business Review* 6 (September): 327–340.

World Bank. 2006. *Doing Business in 2006: Creating Jobs*. Washington, DC: The World Bank.

World Bank. 2007. *Jharkhand: Addressing the Challenges of Inclusive Development*. Washington, DC: The World Bank.

World Bank. 2009. *Doing Business in India 2009*. Washington, DC: The World Bank.

World Bank. 2011. *Doing Business 2011: Making a Difference for Entrepreneurs (Middle East and North Africa)*. Washington, DC: The World Bank.

World Bank. 2013. *Global Financial Development Report: Rethinking the Role of the State in Finance*. Washington, DC: The World Bank.

World Bank. 2015. "World Development Indicators Dataset." Washington, DC: The World Bank.

World Economic Forum. 2013. "The Arab World Competitiveness Report 2013." Insight Report, World Economic Forum and the European Bank for Reconstruction and Development.

World Factbook. 2014. "World Factbook." Washington, DC: Central Intelligence Agency.

Wright, Gavin, and Jesse Czelusta. 2002. "Exorcizing the Resource Curse: Minerals as a Knowledge Industry, Past and Present." Working Paper, Stanford University.

Yadava, Gore Lal. 1981. "Industrialization of Bihar: Some Major Constraints and Possible Solutions." In *Studies in Bihar's Economy and Society*, ed. M. N. Karna, 45–56. New Delhi: Concept.

Yaghoubian, David. 2014. *Ethnicity, Identity, and the Development of Nationalism in Iran*. Syracuse, NY: Syracuse University Press.

Yasin, Kamal Nazer. 2007. "New Era for Iran's Private Sector." Center for Security Studies, ETH Zurich, Switzerland. http://www.isn.ethz.ch/.

Yong, William. 2010. "After Killing at a Bazaar, Iran Declares 2 Days Off." *New York Times*, July 12.

Zahedi, Dariush. 2000. *The Iranian Revolution Then and Now: Indicators of Regime Instability*. Boulder, CO: Westview Press.

Zangeneh, Hamid. 2003. "The Iranian Economy and the Globalization Process." In *Iran Encountering Globalization: Problems and Prospects*, ed. Ali Mohammadi, 107–133. New York: Routledge Curzon.

Zonis, Marvin. 1971. *The Political Elite of Iran*. Princeton, NJ: Princeton University Press.

INDEX

Page numbers followed by *f* and *t* refer to figures and tables respectively. Those followed by n refer to notes, with note number.

Ahmadinejad, Mahmoud, 80, 84, 85, 88
Ahmad Shah (Qajar Shah of Iran), 62
Association for Democratic Reforms (ADR) [India], 118
Australia, non-oil resource production, 11
autocratic governments, mechanisms to guarantee elite rents in, 143–44

Bihar, India. *See also* Undivided Bihar
- bribery and corruption in, 123
- business associations in, 120, 123
- business policies in, 106–14
  - credit availability, 106, 108–12, 114, 122–23, 126, 149
  - focus on SME sector, 113–14
  - lack of distinction between small and large firms, 108
  - reforms, implementation of, 114
  - as small-business-friendly, 103, 106, 114, 126, 149
  - start-up regulations, 106–8, 114, 121
  - and theory of resource windfall effects, 100–102, 106, 108, 128, 129, 150
- creation of, 101, 102*f*, 103, 104
- economic progress of, 127–28
- economy of, bifurcation and, 104–6
- income per capita growth *vs.* Jharkhand, 127, 127*f*
- SME growth *vs.* Jharkhand, 126–27, 126*f*
- SME owners in
  - on business environment, 120–23, 126, 149
  - cooperation between, 121, 123, 126, 149
- as quasi-natural experiment, 101–2

bonyads. See under Islamic Republic of Iran
bribery
- and erosion of government legitimacy, 146
- in Iran, 87
- in Jharkhand *vs.* Bihar, India, 123, 148
- in long-term oil-producers *vs.* other countries, 36–39, 38*t*, 43, 149
- long-term costs of, 39, 146
- in Saudi Arabia, 143

business associations
- bias toward elites in, 53–54
- in Bihar, India, 120, 123
- in Iran
  - under Islamic Republic, weakness of, 89–90, 98
  - under Pahlavi rule, weakness of, 63, 72, 75, 97, 149
  - under Qajar rule, 60–61
- in Jharkhand, India, 103, 120, 123, 125, 149
- oil windfalls and, 45, 52–54, 57
- potential benefits of, 47
- potential leverage of during oil bust periods, 53

business contracts regulation
impact on entrepreneurship, 25
in Iran, 59
in long-term oil-producers *vs.* other countries, 24–25, 24*t*, 149
as function of oil income, statistical test of, 32, 33*t*, 34*t*
measure of, 24
business licensing and permits requirements
in Iran, post-oil period, 59, 68–72, 69*f*, 70*t*, 71*t*, 86, 87, 97
in long-term oil-producers *vs.* other countries, 23–24, 24*t*, 25
as function of oil income, statistical test of, 32, 33*t*, 34*t*
measure of, 24
in Saudi Arabia, 133
business regulation. *See also* business contracts regulation; business licensing and permits requirements; business start-up regulation; business tax regulation; *individual countries*
elite's ability to circumvent, 49
high level of
effects of, 2, 129, 145–47
impact on entrepreneurship, 3, 25, 56–57
and missing middle phenomenon, 3–4
and modernization, 5
and prospects for democracy, 4–5
procedural component of, 22
time component of, 22–23
business regulation reforms. *See also* Saudi Arabia, business regulation reforms in
economic uncertainty and, 147
elites' opposition to, 7, 10, 130, 147
in Gulf Arab oil states, 2, 6, 40, 129, 150, 153
likelihood of, regime stability and, 151
likely political consequences of, 151–52
long-term economic benefits of, 50–51, 148
as low-risk strategy for diversification, 153–54
motives for, 151–52
need for elite rent guarantees prior to, 144
business start-up regulation
in Indian states of Jharkhand *vs.* Bihar, 106–8, 112–13, 114, 121
in Iran, 59
in long-term oil-producers *vs.* other countries, 23, 23*f*, 24*t*, 25, 149
as function of oil income, statistical test of, 30–31, 31*t*, 34*t*
measure of, 23
in Saudi Arabia, reform of, 134
variations in complexity by nation, 22–23
business tax regulation
impact on entrepreneurship, 25
in Islamic Republic of Iran, 59, 86–87, 86*t*
in long-term oil-producers *vs.* other countries, 24, 24*t*, 25, 149
as function of oil income, statistical test of, 32, 33*t*, 34*t*
measure of, 24

Capital Market Authority (CMA) [Saudi Arabia], 134–35
Chamber of Commerce, Industries, and Mines (ICCIM) [Iran], 89–90
Chamber of Commerce, in Iran, 72, 75, 89, 97
Chamber of Industries and Mines (Iran), 72, 97
Coalgate scandal (India), 115–18, 174n117
coal production in India, political impact of. *See also* Jharkhand, India
as test of resource-windfall theory, 100–102, 106, 108, 128, 129, 150, 171n7
as under-researched, 100
Confederation of Indian Industry (CII), 107, 113, 120, 123
contracts regulation. *See* business contracts regulation
controlled comparison tests, 101, 171n7
credit availability
in Bihar *vs.* Jharkhand, India, 106, 108–12, 114, 122–23, 126, 149

crisis of in developing world, 25–26
female entrepreneurship and, 146
government role in, 4, 25
in informal sector, and entrepreneurship, 4
in Iran
as function of oil income, 68, 68*f*
under Islamic Republic, 68*f*, 80, 87–88, 91, 93*f*
under monarchy, 26–28, 26*f*, 27*f*, 66–68, 68*f*, 97
measures of, 26
in oil-producers *vs.* other countries, 26–28, 26*f*, 27*f*, 32–36, 35*t*, 149
in Saudi Arabia, 133–34, 137, 139–40
credit markets, in developing world, 25–26
credit strength
impact of volume of lending, 25
in oil producers *vs.* nonproducers, 27–28
crony capitalism *vs.* economic familism, 132–33
crude oil prices, decline of (1980–2004), 20–21

democracy
and modernization, 5
and oil windfall theory, 101, 129, 150
prospects for, level of regulation and, 4–5
rent-seeking policy makers in, 119
depth of credit information, in oil-producing *vs.* nonproducing countries, 25
developing countries
and cost-benefit analyses for reforms, 50
credit availability crisis in, 25–26
definition of, 159n12
employment in, informal sector and, 4
Dubai, as model of reform, 153
Dutch Disease theory, 5, 7–8, 43

economic familism. *See also* Saudi Arabia, economic familism in
characteristics of system, 132
vs. crony capitalism, 132–33
as substitute for formal institutions, 151
substitutes for in nonmonarchical governments, 143–44
elites. *See also* oil windfalls, and elites; rent seeking; Saudi Arabia, elites in
in autocratic governments, protecting rents for, 143–44
and business regulation, ability to circumvent, 49
definition of, 9
Iranian *bonyads* as functional equivalent of, 78, 81–85, 97, 148
and oil busts, strengthening of bonds to policy makers in, 52
opposition to business regulation reforms, 7, 10, 130, 147
and regulatory capture, 45–46
shaping of policy by, 46
employment
in developing world, informal sector and, 4
in Islamic Republic of Iran
bonyads and, 83–84, 83*f*
decline of SMEs and, 90–92, 90*f*, 91*f*, 93*f*, 95
pro-entrepreneur policies and, 2, 151–52
SMEs and, 3
enclave economies, 11
entrepreneurs. *See also* business start-up regulation; female entrepreneurship; nonelites
as critical factor in development, 3
informal economy and, 4, 146
Iranian bazaar and, 62–63, 73, 91–96, 93*f*–95*f*
in Jharkhand, India, neglect of, 103, 112–14
and level of business regulation, impact of, 3, 25, 56–57
in oil-producing countries, crowding-out of, 8–9, 11, 145
in Saudi Arabia
dependence on royal family, 141–42
potential for political influence, 143, 178n78
variations in political and economic status of, 9

fallacy of unobserved burdens, 152
female entrepreneurship
in oil-producing countries, 146
Saudi Arabia reforms and, 143, 178n76
foreign aid, effects of *vs.* oil windfall, 165n33

government
legitimacy, bribery's erosion of, 146
regime type
and credit availability, 34, 35*t*
and level of business regulation, 29–31, 101
stability of, in oil-producers
impact on credit availability, 34, 35*t*
impact on level of business regulation, 29, 30
Great Britain, in Iran, 61–62
Gulf Arab oil states. *See also* Saudi Arabia
business associations, bias toward elites in, 53–54
business licensing and permits requirements
burden of *vs.* other countries, 23–24, 24*t*
as function of oil income, statistical test of, 32, 33*t*, 34*t*
business regulation
and level of innovation, 41*t*
and new firm entry rates *vs.* other countries, 40, 41*t*
business regulation reforms, recent, 2, 6, 40, 129, 150, 153
business start-up regulation, burden of *vs.* other countries, 23, 23*f*, 24*t*
business tax regulation
burden of *vs.* other countries, 24, 24*t*
as function of oil income, statistical test of, 32, 33*t*, 34*t*
contracts regulation
burden of *vs.* other countries, 24–25, 24*t*
as function of oil income, statistical test of, 33*t*, 34*t*
countries categorized as, 159n7
credit availability *vs.* other countries, 26–28, 26*f*, 27*f*
and "king's dilemma," 143
non-oil income of, 19–20, 20*f*

Hertog, Steffen, 135, 137, 142

ICCIM. *See* Chamber of Commerce, Industries, and Mines
income per capita, in oil-producing countries
and credit availability, 34, 35*t*
and level of business regulation, 29, 30
India. *See also* Bihar, India; Jharkhand, India
business associations in, 47
coal production, and regulation levels
as test of resource-windfall theory, 100–102, 128, 129, 150, 171n7
as under-researched, 100
as democracy, 101
mining regulation in, 117–18
nationalization of coal industry in, 114, 117
policy autonomy of states in, 101
states, bifurcations of, 101–2
informal sector
definition of, 160n15
and employment in developing world, 4
and entrepreneurship, 4, 146
as part of nonelite group, 56
innovation
managerial burden and, 40–43, 41*t*
in oil-producers *vs.* other countries, 40–43, 41*t*, 150
strategies for encouraging, 153
interests, as key to understanding policy, 45–46
Iran. *See also* Islamic Republic of Iran
bazaar in
definition of, 60
and entrepreneurial class, 62–63, 73
Mohammad Reza Shah's campaign against, 73–76, 97, 149
support for revolution, 73, 75–76, 97
White Revolution exclusion of, 67–68, 71–73

White Revolution's economic impact on, 74
business associations in
under Pahlavi rule, 63, 72, 75, 97, 149
under Qajar rule, 60–61
business regulation in
in oil era (1954–1979), 58–60, 97
in pre-oil era, 58–64
and Eurasian trade routes, 61
government institutions, development of, 63
and Great Depression, 63
Law for Protection and Development of Iranian Industries, 82
oil income per capita, 58, 59*f*
Oil Stabilization Fund (OSF), depletion of, 84–85
oil windfall in, 64
economic crisis following, 64–65, 77
effects of *vs.* other factors, 77, 97
post-1954, business regulation in, 58–60, 97
credit availability, 26–28, 26*f*, 27*f*, 66–68, 68*f*, 97, 149
licensing regulation, 59, 68–72, 69*f*, 70*t*, 71*t*, 97
and private sector, development of, 62–64
revolution of 1979
bazaari support for, 73, 75–76, 97
economic uncertainty following, 77
role of unemployed persons in, 152
White Revolution in
antiprofiteering campaign, 74–75
and bazaaris, campaign against, 73–76, 97, 149
business licensing regulation, tightening of, 68–69, 97
centralized policy making in, 72
and concentration of wealth, 68–69
and credit, elite monopolization of, 26–28, 26*f*, 27*f*, 66–68, 68*f*, 97, 149
economic growth under, 65
elite rent seeking in, 65–73, 77, 97
and increase in state monopolies, 66
industrialization in, 65, 74
as modernization campaign, 76–77
and nonelites, increased barriers for, 66–72
oil windfall effects and, 76–77, 97
reforms under, 65
restrictions on Tehran industry, 68
IRGC. *See* Islamic Revolutionary Guards Corps
Islamic Republic of Iran
anti-capitalism in, 78, 88, 89
banking reforms under Khatami, 80
bazaar in
decline of entrepreneurship in, 91–96, 93*f*–95*f*
and government anti-capitalism, 88, 89
heavy regulation of, 87, 98
political weakness of, 88, 89, 98
as source of financing, 87
status of, 78–79
strikes and protests by, 88–89
and wholesale trade, *bonyads*' activities in, 93–95, 94*f*
bonyads in
access to credit, 87
and anticapitalist attitudes in Iran, 78
autonomy from government, 81–83
economic inefficiency of, 169n138
functional similarity to private-sector elites, 78, 81–85, 97, 148
government funding of, 81
growth of, 83–84, 84*f*, 85
increased activity under Ahmadinejad, 80
increasing trade sector activity, 93–95, 94*f*, 95*f*
percentage of non-oil economy controlled by, 82
and privatization, benefits from, 85, 98
rent seeking by, 81, 82, 97
resistance to reforms, 83, 130
as source of manufacturing growth, 91, 92*f*
support for government, 84, 148
transfer of seized elites' assets to, 82
welfare activities by, 81, 82

Islamic Republic of Iran (*Cont.*)
business associations, weakness of, 89–90, 98
business regulation in
business start-up regulation in, 59
as continuation of pre-revolution policies, 60, 77, 97
as heavy and complex, 85–88, 95, 98
impact on private sector growth, 90–96, 98
international sanctions as explanation for, 96
licensing and permits requirements, 86, 87
as product of oil windfall, 88, 96, 98–99
statism as explanation for, 96
credit availability, 68*f*, 80, 87–88, 91, 93*f*
economic policies, haphazard nature of, 96
employment in, SME decline and, 90–92, 90*f*, 91*f*, 93*f*, 95
IRGC in
functional similarity to private-sector elites, 84–85
increased activity under Ahmadinejad, 80
and militarization of economy, 169n139
manufacturing sector
bonyads and public firms in, 91, 92*f*
credit availability for, 91, 93*f*
decline of nonelite entrepreneurship in, 94–95
growth of, 91, 92*f*
nationalization of banks in, 80
nonelites, business environment for
credit availability, 80, 87–88, 91, 93*f*
and decline in entrepreneurship, 90–96, 98
high levels of regulation, 85–88, 95–96, 98
as product of oil windfall, 88, 96, 98–99
tax burden, 86–87, 86*t*
nonelites, political weakness of, 88–90, 98
oil bust, repression of nonoil sector growth in, 91, 98
oil income per capita, 19*t*, 58, 59*f*
price controls in, 86
and private property, as concept, 79
private sector in
debate on role of, 78–80
investment decline in, 79
privatization of state-owned enterprises
bonyads and IRGC as primary beneficiaries of, 85, 98
debate on, 80–81
slow pace of, 85
redistribution of wealth in, 77–79
SMEs in
decline of, 1, 90–92, 90*f*, 91*f*, 93*f*, 95
importance to economy, 85
political weakening of, 85
state intervention in economy in, 79
subsidy reforms, delays in, 89–90
tax burden, on SMEs *vs.* other firms, 86–87, 86*t*
trade sector
bonyads' increasing role in, 93–95, 94*f*, 95*f*
decline of entrepreneurship in, 91–96, 93*f*–95*f*
war with Iraq, and state intervention in economy, 79
Islamic Revolutionary Guards Corps (IRGC)
functional similarity to private sector elites, 84–85
increased activity under Ahmadinejad, 80
and militarization of economy, 169n139

Jharkhand, India. *See also* Undivided Bihar
bribery and corruption in, 123, 148
business associations in, 103, 120, 123, 125, 149
business policies in, 106–14
corruption in, 115–20
credit availability, 106, 108–12, 114, 122

disadvantaging of small firms, 108, 111, 114
government instability and, 123–25
Industrial Policy documents, 112–13
mining sector as focus of, 105, 112, 118–20, 125
neglect of local entrepreneurs in, 103, 112–14
as quasi-natural experiment, 101–2
reforms, lack of follow-through on, 112–13
start-up regulations, 106–8, 112–13, 121
and theory of resource windfall effects, 100–102, 106, 108, 128, 129, 150
coal mafia in, 123, 124–26, 148
creation of, 101, 102*f*, 103, 104
development plans, 104
economic future of, 128
government instability in, 119, 120, 123–26
income per capita growth, 127, 127*f*
industrial advantage over Bihar, 106
mining-dependent industries in, 114–15
and mining firms, rent seeking by, 114–20, 125, 148
mining in
Coalgate scandal, 115–18, 174n117
government focus on, 105, 112, 118–20, 148
government royalties from, 105, 105*f*
illegal coal trade and, 124–26
as issue in independence movement, 104
mining lease distribution, 114–20
as percent of gross state domestic product, 100
Naxalite insurgency in, 123, 124
policy makers
coal mafia and, 125, 148
necessity of courting coal industry, 118–20, 125
political implications of supporting SMEs, 119, 120
rent seeking by, 103, 114, 116–20, 125, 148
and resource curse, 127
restrictions on entrepreneurs and small firms, 103
SME growth *vs.* Bihar, 126–27, 126*f*
SME owners in
on business environment, 120–23, 149
conflict among, 121–23, 125, 149

Karl, Terry Lynn, 8, 36, 162n1
Khamenei, Ayatollah, 79, 80, 96
Khatami, Seyed Mohammad, 79–80, 84, 98
Khomeini, Ayatollah, 78, 79
"king's dilemma," Saudi Arabian reforms and, 143

managerial burden
on long-term oil-producers *vs.* other countries, 36–39, 38*t*, 43, 149
and rate of innovation, 43
MENA region. *See* Middle East/North Africa (MENA) region
microfinance movement, 25–26
middle class, growth of, and prospects for democracy, 4–5
Middle East/North Africa (MENA) region
business regulation
and level of innovation, 41*t*, 42
managerial burden of *vs.* other countries, 36–37, 38*t*
and new firm entry rates *vs.* other countries, 39, 40, 41*t*
contracts regulation, 24–25, 24*t*, 33*t*, 34*t*
level of bribery *vs.* other countries, 37, 38*t*
level of credit availability as function of oil income, 34, 35*t*
licensing and permits requirements, 32, 33*t*, 34*t*
monarchies in, 131, 151
start-up regulations, 29–31, 31*t*, 34*t*
tax regulation, 32, 33*t*, 34*t*
missing middle phenomenon, 3–4

modernization
as contested concept, 143
definition of, 5
and "king's dilemma," 143
level of business regulation and, 5
Mohammad Reza Shah (Shah of Iran). *See also* Iran, White Revolution in
and antiprofiteering campaign, 74–75
business environment under, 64
oil windfall *vs.* other explanations for, 76–77, 97
campaign against bazaaris, 73–76, 97, 149
centralization of policy making under, 72, 79
deposing of, 77
and elite rent seeking, support for, 66, 72–73, 77–78, 148
oil windfall under, 64
political instability under, 76
and *Rastakhiz* party, 74–75
and SAVAK, 75
Mohammad Shah (Qajar Shah of Iran), 61
monarchies
and "king's dilemma," 143
in MENA region, 131, 151
stability of, and ability to implement reforms, 131–32

natural resource abundance
and rent seeking, 7
and state capacity, 50
Naxalites, in Jharkhand, India, 123, 124
new firm entry rates, in oil-producers *vs.* other countries, 39–40, 41*t*, 149–50
as function of oil income, statistical test of, 40, 41*t*
nonelites. *See also* business regulation; entrepreneurs; informal sector; small- and medium-sized enterprises (SMEs)
definition of, 2, 9, 56
freedom of, and prospects for democracy, 4–5
and modernization, 5
post-oil windfall decrease in political power, 11, 44–45, 47, 48*f*, 53–54, 147–49
diversity of as political handicap, 52, 53
and increased chance of punishment, 45, 54, 57
in Iran, 60, 73–76
in Islamic Republic of Iran, 88–90, 98
policy makers' interests and, 51–52
non-oil-producing countries, income of *vs.* non-oil sectors of oil-producing countries, 18–21, 20*f*, 43
non-oil sectors. *See under* oil-producing countries

oil, as term, 159n5
oil boom era (1973–1985), and oil-producing countries
credit availability in, 26, 26*f*, 27*f*, 34–36, 35*t*
and non-oil sector income, 19–20, 20*f*, 43
oil bust era (1986–2003), and oil-producing countries, 34–36, 35*t*
credit availability in, 26–27, 26*f*, 27*f*
and non-oil sector income, 20, 20*f*
oil busts
and bond between policy makers and elites, 52
and rent distribution, efforts to reduce, 49
oil-producing countries. *See also* oil windfalls
authoritarianism of, rentier state theory on, 5
business licensing and permits requirements
burden of *vs.* other countries, 23–24, 24*t*, 25
as function of oil income, statistical test of, 32, 33*t*, 34*t*
business regulation in, 1–2
burden of *vs.* other countries, 21–25, 23*f*, 24*t*, 149
as function of oil income, statistical test of, 29–36, 30*t*, 31*t*, 33*t*, 34*t*, 40, 41*t*
as independent of other variables, 2, 6
lack of support for entrepreneurs in, 8–9, 11, 145

and level of innovation, 40–43, 41*t*, 150
as limit on modernization, 5
managerial burden of *vs.* other countries, 36–39, 38*t*, 43, 149
and new firm entry rates *vs.* other countries, 39–40, 41*t*, 149–50
rentier state theory and, 6–7
business start-up regulation
burden of *vs.* other countries, 23, 23*f*, 24*t*, 25, 149
as function of oil income, statistical test of, 30–31, 31*t*, 34*t*
business tax regulation
burden of *vs.* other countries, 24, 24*t*, 25, 149
as function of oil income, statistical test of, 32, 33*t*, 34*t*
contracts regulation
burden of *vs.* other countries, 24–25, 24*t*, 149
as function of oil income, statistical test of, 32, 33*t*, 34*t*
credit availability
as function of oil income, statistical test of, 32–36, 35*t*
vs. other countries, 26–28, 26*f*, 27*f*
credit strength *vs.* nonproducers, 27–28
depth of credit information *vs.* other countries, 28
diversifying of economy, strategies for, 153–54
domestic benefits funded by oil, 18
Dutch Disease theory of, 5, 7–8, 43
economic growth rates *vs.* other countries, 159n6
and economic shocks, vulnerability to, 26
economic volatility of, 159n6
female entrepreneurship in, obstacles faced by, 146
governments of
control of rent seizing, 11
focus on elite rents, 145
reduced incentives for business reforms, 10–11
and innovation, strategies for encouraging, 153
level of bribery *vs.* other countries, 36–39, 38*t*, 43, 149
long-term, countries identified as, 18, 19*t*
non-oil sectors
benefits of expanding, 17–18
factors in stagnation of, 21
importance of growth in, 21
income, calculation of, 18, 162n5
income of *vs.* other countries, 18–21, 20*f*, 43
money potentially available to, 21
stagnation of, and weak economic performance, 1
strategies for avoiding stagnation in, 152–53
as under-researched, 18
oil income per capita, 19*t*
and oil sector, focus on, 145
policy making logic in, 12, 44
rentier state theory of, 5–7, 43, 87
taxation rates, rentier state theory and, 6
weak economic performance, 1, 145
oil windfalls. *See also* rent seeking
burden of managing, 152–53
and business associations, 45, 52–54, 57
and elites
desire to eliminate competition, 10, 46, 48–49
greater influence on policy makers, 47
increased rent seeking by, 9–12, 48, 48*f*, 147–48
increased resistance to reform, 44–45, 49
isolated support for nonelites among, 165n16
policy maker incentive for supporting, 47, 49, 51, 56, 147
types of rents sought by, 48
and Islamic Republic of Iran business regulation, 88, 96, 98–99
and nonelites decrease in political power, 44–45, 47, 48*f*, 53–54, 147–49
diversity of, as political handicap, 52, 53
in Iran, 60, 73–76
in Islamic Republic of Iran, 88–90, 98
policy makers' limited incentives to accommodate, 45, 51–52, 54, 57

oil windfalls (*Cont.*)
punishment of collective action by, 45, 54, 57
and policy makers
decreased incentives for reform, 44–45, 48*f*, 50–52, 147
decreased likelihood of supporting nonelites, 45, 51–52, 54, 57
incentives for providing rents to elites, 47, 49, 51, 56, 147
and regime stability, impact on, 151
testing of hypotheses about, 54–56

Pahlavi dynasty. *See* Mohammad Reza Shah; Reza Shah (Shah of Iran)
policy makers
and cost-benefit analyses for reforms, 50–52, 148
in Jharkhand, India
coal mafia and, 125, 148
necessity of courting coal industry, 118–20, 125
political implications of supporting SMEs, 119, 120
rent seeking by, 103, 114, 116–20, 125, 148
and oil windfalls
decreased incentives for reform, 44–45, 48*f*, 50–52, 147
decreased likelihood of supporting nonelites, 45, 51–52, 54, 57
incentives for providing rents to elites, 47, 49, 51, 56, 147
and rent distribution, motives for reducing, 49
rent seeking by, 46, 56
in Jharkhand, India, 103, 114, 116–20, 148
and rent seizing, desire to maintain, 10–11, 51, 52, 115, 148
policy-making environment
capture of by business interests, 45–46
shaping of by capacity of agents, 46–47
shaping of by interests, 45–46
shaping of by rent seeking, 46
policy recommendations on business environment, 9, 152–54
political donations, 118, 148
process tracing, 13–14, 161n65

Qajar dynasty (Iran)
business environment under, 60–62, 64
European quasi-colonial rule and, 61–62, 64
decentralized administration of, 61
overthrow of, 62

Rafsanjani, Ayatollah, 79, 98
reforms. *See* business regulation reforms
regulatory capture, 45–46
rentier state theory, 5–7, 43, 87
rents, definition of, 7
rent seeking
decreased competition as motive for, 10, 46, 48–49
definition of, 7
higher frequency of in natural resource-based economies, 7
increased profits as motive for, 10, 48
in Iran
under Islamic Republic, 81, 82, 97
under monarchy, 60, 65–73, 77, 97, 129
in Jharkhand, India
by coal mafia, 125
by mining firms, 114–20, 148
by policy makers, 103, 114, 116–20, 148
non-oil windfalls and, 11–12
in oil-producing countries, 9–10
and elite opposition to business reforms, 7, 10, 130, 147
oil windfalls and, 9–12, 48, 48*f*, 147–48
and political stability, 7
shaping of policy by, 46
types of rents sought, 46, 48
rent seizing, 10–11, 51–52, 114–15, 132, 148, 161n59
resource curse. *See also* Dutch Disease theory; rentier state theory
Jharkhand, India, and, 104, 127
scholarship on, 9, 17
resource windfalls, non-oil
avenues for research on, 150
income from *vs.* oil windfalls, 11
similarity to oil windfalls, 11–12, 103, 129

Indian coal-producing states as test of, 100–102, 106, 108, 128, 129, 150
Reza Shah (Shah of Iran)
abdication of, 64
business environment under, 62–64, 97
overthrow of Qajars by, 62
support for bazaaris, 73
Ross, Michael, 18, 119, 146, 152, 159n6
Russia, 1, 19*t*, 61

SAGIA. *See* Saudi Arabian General Investment Authority
Al-Sanea family, 139
Saudi Arabia
blurred line between politics and finances in, 140–41
bribery in, 143
business regulation reforms in, 129, 133–37, 150–51
evaluations of success of, 134
factors motivating/allowing, 130–31
new institutional bodies, 133–34
support for start-ups, 134–35
work remaining to be done, 136–37, 143
credit availability in, 133–34, 137, 139–40
economic familism in
characteristics of, 132
and credit availability, 139–40
vs. crony capitalism, 132–33
development of, 137–40
and freedom to pursue reforms, 130–32, 143, 151
influence on Saudi bureaucracy, 142
intermarriage and, 137–38
and lack of transparency, 139
and royal family business elites, 140–42
royal support for, 139, 140
as stabilizing royal-elite bond, 132, 142
elites in
crowding out of newcomers by, 139
dominance of private sector, 139
hands-off policy toward, 139
intermarriage, 138–39
protected rents of, and freedom to pursue reforms, 130–32, 139, 140, 142, 143
rise of elite families, 137–38
and entrepreneurs, potential for political influence, 143, 178n78
and "king's dilemma," 143
monarchical system
need for binding commitment to elites, 131
stability of, and freedom to pursue reforms, 130–31, 151
and oil bust period, efforts to reduce rents during, 49, 52
oil income per capita, 19*t*, 130
private non-oil sector, growth of, 135–36, 135*f*, 136*f*
privatization in, 135
as proxy for developing states, 130
royal family
business elites in, and economic familism, 140–42
nonroyal entrepreneurs dependent on, 141–42
support for economic familism, 139, 140
and World Trade Organization, 133
Saudi Arabian General Investment Authority (SAGIA), 133–34
Saudi Industrial Development Fund, 137
SAVAK, 75
SEC. *See* Supreme Economic Council (SEC) [Saudi Arabia]
Sepah e Pasdaran-e Enqelab-e Islami. *See* Islamic Revolutionary Guards Corps (IRGC)
small- and medium-sized enterprises (SMEs). *See also* nonelites
definition of, 159n13
and entrepreneurship, 3
as focus of this study, 2
growth Bihar *vs.* Jharkhand, India, 126–27, 126*f*
growth opportunities, restrictive business environment and, 147
influence of, factors affecting, 47
innovative firms as, 42
in Islamic Republic of Iran

small- and medium-sized enterprises (SMEs) (*Cont.*)
decline of, 1, 90–92, 90*f*, 91*f*, 93*f*, 95
importance to economy, 85
political weakening of, 85
tax burden on, 86–87, 86*t*
and missing middle phenomenon, 3–4
in oil-producing countries
economic importance of, 2–3
limited research on, 2
as suppliers for larger firms, 147
stability of government
and credit availability, 34, 35*t*
and level of business regulation, 29, 30
and likelihood of reforms, 6–7, 151
rent seeking and, 7
state capacity, natural resource abundance and, 50
stationary bandit theory, 144
statistical tests of regulation measures, methodology, 28–29, 30*t*, 163n32
Supreme Economic Council (SEC) [Saudi Arabia], 133, 142

Tata Iron and Steel Company (TISCO), 105, 115, 117–18
trade, localized, transition to small-scale manufacturing
and economic development, 94
in Islamic Republic of Iran, 94–95, 95*f*

Undivided Bihar. *See also* Bihar, India; Jharkhand, India
coal mafia in, 124, 125
division of, 101, 103, 104
large-scale mining investment in, 115
resource curse in, 104
socioeconomic characteristics, 103–4

windfalls. *See* oil windfalls; resource windfalls, non-oil
women. *See* female entrepreneurship
World Bank
data on Iranian tax burden, 86
Doing Business project, 21–22
Doing Business in India, 106–8, 121
government efforts to manipulate, 163n32
on MENA business environments, 151, 162n9, 163n32
on Saudi Arabia reforms, 134
Enterprise Surveys data, 36, 37, 162n9
on Jharkhand, 112, 127
on Saudi Arabia, 129–30
Worldwide Governance Indicators, 134

www.ingramcontent.com/pod-product-compliance
Ingram Content Group UK Ltd.
Pitfield, Milton Keynes, MK11 3LW, UK
UKHW021134160926
475595UK00020B/39

9 780190 490218